Solving Plant Problems
Design, Operation, Maintenance

William O'Keefe
and
Thomas C. Elliott, Editors

McGraw-Hill, Inc.   New York

Library of Congress Cataloging in Publication Data

Main entry under title:

Solving plant problems.

    Reprint of items originally published in Power, 1973-
1984.
    Includes index.
    1. Power-plants.   I. O'Keefe, William.   II. Elliott,
Thomas C.
TJ405.S68   1984   621.31'2132   84-12605

ISBN 0-07-606905-2

Preface

This collection of problems and solutions from the pages of POWER represents the best material from "This Month's Problem" (formerly "Plant System Problems")—a reader participation department in the magazine. The 75 items in the collection go back over 11 years, to January 1973, but are just as applicable today as when they were written; that is, the problems still exist, and the solutions still are viable.

As with POWER readers, the readers of this book are most likely to be engineers in electric utilities, process and manufacturing industries, commercial and service establishments, plus consulting engineers working in the power technology. The book should also appeal to engineering students and engineers with vendors in the power field.

The book is divided into eight sections that roughly parallel the long-time interest areas of the magazine. Each section starts with items published recently in POWER and moves backward chronologically to older items, because it was easier to select them on that basis. It's a small point, however, because of their timeless quality.

Each of the 75 numbered items herein presents a problem followed by a range of solutions, usually five to a dozen, depending on problem complexity and individuality of the answers. In a particular issue of the magazine, a problem would be (and is) posed to readers, inviting a response and announcing that publishable answers would appear in an issue four months later. That month's solutions in the issue would be (and are) in response to a problem posed four months earlier.

You should find the Table of Contents a big help—all 75 problems are identified, section by section. For detail queries, the 800-entry index should be invaluable in pinpointing systems, equipment, and techniques.

If you have a problem or two in your plant like those described in this book, the solutions may well lie in these pages. Or you may just wish to match wits with our contributors in your specialty—like boilers, corrosion control, or pipe systems. Whatever your reasons, however, the experience should be rewarding. The 75 problems, and many more solutions, touch on all aspects of the power field.

WILLIAM O'KEEFE AND THOMAS C ELLIOTT, EDITORS

Table of Contents

7. Energy Management

8. Plant Systems and Equipment

1 Steam Generation

1.1 Sudden boiler-fan failures: What to do

A recent accident with an induced-draft fan on one of our larger boilers, a 500,000-lb/hr pulverized-coal unit, has made us consider steps to prevent recurrence of the loss. What happened was that the fan flew apart practically without warning, wrecking the housing and some ductwork on an adjacent boiler breeching, plus doing some damage to the bearings of the drive motor.

Our fan supplier explained it as mostly something that happens every once in a while but also said imbalance from deposits was partly responsible. For the replacement, we installed vibration-indicating gages that are checked twice a shift, but I am not sure that this is enough. I realize that we, as a small plant compared with big central stations, cannot go as far in quality control or on-line instrumentation as the owners of large boilers. Nevertheless, we would like to get better reliability from boiler fans. What has been the experience of POWER readers with specifying for better fans? What is reasonable instrumentation for detecting trouble in a running fan? Should we dye-check the welds or certain other parts like shafts periodically? Would ultrasonic examination by an outside agency, say, twice a year, be cost-effective? — KDS

Inspect for specific trouble sources

Conditions working to the detriment of an induced-draft fan are: corrosion, erosion, dirt accumulation, and heat. These conditions interact, and operators must be aware of this and attempt to control the conditions. Visual inspection at each maintenance or extended-outage period should be directed to shaft corrosion because of "dewing" of flue-gas condensables, although this is usually a concern only for double-suction fans, in which shaft-penetration points through the casing are below atmospheric pressure and let cool outside air in. Look for dirt and ash accumulations on fan blades, too, and see if the buildup is uniform or if there are spots where flaking has occured. Erosion can happen, but pulverized-coal applications are usually the victims. Check suction dampers, as well.

All of these items can cause vibration or cracking at welds. For weld test, make a dye-penetrant check every three to five years, depending on the ash-

1

deposition experience. A trained technician should interpret the developed dye test. Heat-treating the rotating assembly during manufacture will reduce the stress points and corrosion potential. Proper maintenance includes painting of shafts where corrosion has occurred, say, at air inleakage points. Shaft seals, properly maintained, will reduce inleakage. Annually, check and maintain balance of the rotating assembly. Clean off fan-blade deposits, especially before balance check.

Adjusting burners to operate at a low excess-air level will reduce SO_2-to-SO_3 conversion and depress dewpoint of the flue gas. Also, keep boiler casings tight to prevent air inleakage, reducing the SO_2-to-SO_3 conversion. Both these measures will keep ash dry and friable and, with a mag-oxide based fuel treatment, will reduce ash accumulation and subsequent corrosion. For monitoring, proximity probe monitors are low- to moderate-cost devices that give easily interpreted and continuous monitoring of the i-d fan. They can give warning alarms and trip at critical levels. A periodic check with accelerometer pickups will give data on acceleration (g), peak frequencies, and amplitudes over a spectrum. And graphs give a historic data base for later comparison.

M J Fox, *Fort Edward, NY*

Signs will always presage failure

Determining the cause of failure is the first step, before investing in a lot of expensive and unwarranted instrumentation. (The vibration-indicating equipment is a good, idea, however.) Several causes could have been working in KDS's trouble. Ash deposits, or corrosion from ash abrasiveness, could have given rotor imbalance. Drive-coupling misalignment (assuming babbitt sleeve bearings), inadequate oil clearance causing high temperatures and wiped bearings, or excessive oil clearance that will cause oil swirl are other reasons for fans to fly into pieces.

Try preventive maintenance here, with handheld vibration-monitoring equipment on horizontal, vertical, and axial motion of fan and motor. Motor-amp load with fan loaded and unloaded should be taken. Temperature of bearings and equipment and periodic inspection of fan rotor for deposits and corrosion are important. Weekly inspection of oil-slinger rings, if they are accessible, is advisable. Unless there are strong suspicions, ultrasonic exam is unnecessary. I have yet to see a piece of equipment fail suddenly and without signs of the pending failure. Such signs are excessive noise, above-normal vibration or bearing temperature, and high motor load.

R A JOHNSON, *Snowflake, Ariz*

Vibration activates motor-cutoff switch

The installed vibration gages would be satisfactory, but KDS should add a motor-cutoff switch actuated by severe vibration. This will prevent damage, not merely indicate it. Periodic dye check is unnecessary if there is initial certification on critical requirements. If necessary at all, dye-check on a new fan before acceptance or on a fan that has failed from external causes. Ultrasonic is probably not cost-effective. Specific failures that are caused externally include foreign objects drawn into the air stream. Low-resistance screens, checked periodically for blockage, can help here, as can guard rails around the housing openings. Loosening of bolts and other restraints under temperature change can also make trouble. Apply locking compounds to all assembly fits.

G O Trias, *Rockford, Ill*

Set vibration alarm to control room

Deposits on the blade and bearing wear are two common causes of blade failure. Material ordinarily deposits equally on all blades, but minor differences in deposit composition or blade surface cause clumps of material to fly off and increase both vibration and bearing wear. Continued buildup and de-clumping increases both vibration and metal stress until the blades fracture. Analysis of deposits will aid in source identification. Install bearing vibration pickups with local indicators, connecting the pickups to a control-room recorder with high-level alarms. Get recommended settings from the fan manufacturer.

J E Janda, *Hancocks Bridge, NJ*

Build reliability into specifications

KDS should have started back during procurement, by adding a prequalifying clause, such as "Bidder must have successful operating experience on similar-capacity or larger fans in similar applications in at least three different installations, each logging a minimum 10,000 hours of accumulated run at the date of bidding." In specifying, set out conservative operating conditions for fan inlet, with a minimum of 27 deg F over maximum expected operating temperature of flue gas and dust loading of 0.15 gr/sft^3, the latter condition taking into account partial outage of dust-collecting apparatus.

It might be better to specify a backward/radial-plate-type double-inlet centrifugal instead of a more efficient airfoil-blade type, too, because on-site repair, with welding and static balancing, is easier. A variable-speed i-d fan instead of a fixed-speed one with inlet-vane-angle control could be better on erosive coal-fired service, too.

Specify limit of 0.1 in./sec velocity peak in vibration. For the fan speed, a conservative upper limit could be 900 rpm, calling for an eight-pole motor. Quality-assurance measures for the specification could include radiography on impeller welds, dye-penetrant test on other welds, ultrasonic on the shaft, and static and dynamic balance on the finished rotor. In operation, with a proximity-probe sensor, a fan can run continuously without problem in the upper zone of the specified limit, but any shift in vibration magnitude, even well within the limit, is a sure indicator of problem onset. Investigate it promptly.

B K Mitra, *Philadelphia, Penn*

Strobe light is on-line inspection tool

KDS might try replacing the inspection doors with clear panels for inspection of the fan internal elements. A strobotac instrument is useful then for a check on such elements as blading, struts, keyways, or builtup deposits, while the fan is running. Remember, too, that a major cause of fan accidents is torsional stress produced by bearing overheating and friction rub in the fan casing.

E R GARLOCK, *Deerfield, Ohio*

1.2 Why coal-burner burn out?

One of our pulverized-coal-fired units has recently experienced burned-out PC burners, after running satisfactorily for a long time. We have checked over conditions and settings, but cannot find any change in operation beyond a change in coal from our usual high-sulfur Illinois coal to a Kentucky coal. The Kentucky coal has 2000 Btu/lb more heat than the Illinois coal. Ash is 2% lower, but grindability and volatiles are alike.

The stockpiled Illinois coal had considerably more dust fines (up to 10% of total) and increasingly high ash content as we neared the bottom of the pile, but these changes had not given us any trouble before, even though we had had to do more grinding to stop slagging in the furnace. Can POWER readers suggest steps to solve this trouble—without making large-scale changes in burners or switching fuel, which would be out of the question for a long time? We need to be able to burn the Kentucky coal to keep costs down.—LI

Adjust ignition by damper setting

Sounds as if LI's Kentucky coal is igniting too close to the nozzles. Every type of coal has its own peculiarities and must be fired differently. When we were firing coal some 30 years ago, our corner-fired boilers were new to us. Our older boilers were front-fired and had their own primary-air fans. The new boilers had forced-draft fans and windboxes with damper settings behind the coal stream, as well as above and below. Mill exhausters blew the coal into the boiler as primary air, and dampers on the windboxes admitted secondary air.

Observation of the fire through inspection doors above the firing level enabled us to regulate the point at which ignition of the coal took place by adjusting the damper behind the coal—with best results from ignition about 16 in. from the burner. By frequent checking, we established damper settings suitable for the coal that we were burning. Two observers, each checking firing conditions twice per shift, gave us good control, and we had no problems of burning out corners. Besides, the frequent checking prevented slagging over of burner nozzles, because slag could be quickly seen and rodded off.

U SYDOW, *Glenwood Landing, NY*

Coarser grind moved flame out from burner

In a similar case, we realized that we had to move the flame out from the burner. We checked the fuel velocity and found that it was normal—80 ft/sec at high load. Check of fineness indicated that 86% went through 200-mesh screen, which indicated that the coal was too fine. Therefore, we adjusted for a coarser grind, and almost immediately the flame moved out from the burner to where we wanted it. We had realized that we needed to decrease the flame velocity. The coarser coal would burn farther out from the burner because of the longer time needed.

A CHAO, *Jacksonville, Ill*

Check all settings by observation

Changing from a high-sulfur Illinois coal to a Kentucky coal with 2000 Btu/lb more heating value would ordinarily call for changes in burner settings and very possibly in pulverizer classifier settings. Apparently, LI had not made the necessary changes when fuel was changed. Burners should be adjusted to get the "best fire," which is a fire that:

■ Does not burn in too close to the burner; if it does, it will overheat the burner tip and eventually destroy the burner.

■ Does not impinge on the furnace walls.

■ Does not have sparklers from coarse coal.

■ Is bright but not too bright, indicating that the excess air is in the right range.

Learning how to get the "best fire" calls for observation of fires and of what happens when changes come. Hands-on training, under an experienced teacher, is necessary. Trained personnel should check burners after every large load change and whenever there is a change in fuel. On large boilers, visual check is more difficult but no less important than on small boilers. Check of settings alone is risky. Here's why:

■ Linkages can come loose, allowing regulators to close or open, even with the correct setting.

■ Couplings in control drives break, giving false indications in the control room.

■ Pointers come loose and then don't indicate actual positions, resulting in errors going unnoticed.

J L RUSS, *Prairie Village, Kan*

Overheated spreaders make trouble

LI doesn't give enough details of his operation to allow really pinning down the problem, but one thing to look at is pulverizer-discharge temperature. Some coals will coke on the coal spreaders at high temperatures and eventually allow the spreader to overheat and burn out. LI might try reducing mill-discharge temperatures in 10-deg F increments. We went to the trouble of installing thermocouples down through the center of the spreader support tube to measure the temperature near the furnace end. This temperature increases rapidly if caking becomes extensive.

An alarm set at 250F warns our operators of the condition. Then a couple of light blows from a small sledge will usually clear caked coal from the spreaders and let the primary-air/coal mixture cool the spreaders again. We cool spreaders to below the alarm point with primary air alone, before putting coal into the pulverizer. This prevents the hot spreader, which can be over 100F in our case, from baking the initial surge of coal onto itself.

J R Cimini, *New Martinsville, WVa*

Car-seal straw can block cooling-air flow

Because the Kentucky coal burned now has a larger heating value than the previous fuel, there may be a higher radiant-heat temperature at the burners, and pulverized coal may coke at burner tips. Ordinarily, air/coal-flow rates are enough to cool the burner tips enough to prevent coking. If steam load drops, however, the lowered air-flow rate may not give enough cooling.

Another possible source of difficulty is the straw that often is put into the bottom of hopper cars to seal the gates. The straw can carry through the coal-storage and grinding circuits and then pack behind burner rosettes. The resulting reduction in cooling-air flow may cause the burner to burn out. In any case, high-alloy steels for burner surfaces exposed to high temperatures are recommended to increase equipment life.

H B Wayne, *Jamaica, NY*

Is coal-pipe velocity high enough?

Two important areas need consideration when trying to determine the causes of burner deterioration in a pulverized-coal boiler. LI should first examine the retractable burner impeller to make sure that it does not extend into the area of actual combustion when in the coal-firing position. If any adjustment is necessary, limit impeller travel to that specified by the manufacturer. Second, and most important, is the subject of coal-pipe velocity and primary-air flow. When the boiler was started up, the boiler control system was adjusted so that the correct fuel/air ratio would be maintained for the specified coal analysis. Both primary and secondary air were taken into account.

In addition, provisions assured that the velocity in the coal pipe stayed at or above a safe value, typically 3000 ft/min. There are many reasons why the control system could be out of adjustment. In LI's case, perhaps some adjustments were made to improve combustion of the poor-quality coal burned during conversion to Kentucky coal. In addition to the burner-deterioration problems, an improper fuel/air ratio can give an explosive, fuel-rich mixture in the boiler. And finally, inadequate coal-pipe velocity, especially in a low-firing condition, can result in a potentially serious coal-pipe fire.

F P Newman, *Dublin, Ga*

ength of shutdown contribute to the problem? If we have to shut down for long periods again, how can we prevent recurrence?—IGK

Continuously operating unit is OK

Investigation would show that cracking in the condenser tubes resulted from stress corrosion, under simultaneous action of a sustained stress and a specific corrosive environment. The stress can be low and the corrosive environment mild, but the combination gives sudden unexpected failure. The stresses can be residual in the metal, applied tensile stress, or cyclic stress. Exposure of brass to moist atmosphere with traces of ammonia is especially drastic. Ammonia and its compounds are believed to be the potent agents. CO_2 may contribute.

In a similiar case that we experienced, we had no failures whatsoever on an identical unit in continuous operation for longer than the troublesome condenser. Cleaning the tubes thoroughly and keeping them dry could probably have prevented our and IGK's mishaps.

S NASRULLAH, *Guddu, Pakistan*

Remove at least one causative factor

The stress-corrosion mechansim is a time-dependent one, with specific factors: a susceptible material, tensile stress, corrosive environment with agents such as caustics, chlorides, sulfides, ammonia, or mercury. IGK would have to remove one or more of these factors to prevent stress-corrosion cracking. In the absence of information on location of the failures (whether at the tubesheet periphery, in the air-removal section, scattered, or elsewhere), there are several possible answers.

If failures are in the peripheral area of the condenser tubesheet, then axial force from water pressure on the waterboxes could be tensioning the tubes. Catenary sag between tube-support plates could also stress the tubes. The tubes should have enough slope to be self-draining. From my experience, however, many tubes cannot drain because of warpage or blockage. If so, then IGK's shutdown procedure should try to reduce problems caused by stagnant water trapped in the tubes. One procedure is to maintain circulation through the condenser when no work is being done on it. Also, blow out the tubes with air when periods exceeding three days of no circulation are anticipated, and follow by clean-water flush and reblowing with air.

Agents on the steam side may be the trouble. If most of the failures are in the air-removal section, then ammonia-induced stress corrosion could be a possibility. Last, cyclic vibration, from inadequate tube-support spacing, poor tube fit in supports, adjacent rotating machinery, or steam entering the condenser could be causes of cracks. Best solution to all stress-corrosion problems is to retube with a more resistant metal, such as 90/10 or 70/30 copper/nickel, or perhaps 304 stainless steel.

J A SCHMULEN, *Fort Worth, Tex*

Try trace injection of ferrous sulfate

If the water analysis is representative, then the water is slightly on the corrosive side, with Langelier index of 0 to +0.4 and Ryzner index of 7.3 to 7.8. Tube and tubesheet materials are questionable, because copper-bearing alloys have proven susceptible to corrosion by sulfide ions. IGK's water has sulfate, and in the presence of sulfate-reducing microorganisms, severe corrosion is to be expected. Although change of tubes to 304 stainless would be ideal, a simpler and cheaper fix could be trace injection of iron ions (0.05 ppm of ferrous sulfate, say) ahead of the condenser to increase corrosion resistance.

Although IGK gives no details on the method for sealing the unit to prevent ingress of air into the condenser from water and steam sides during the long shutdown, there should be no special corrosion problems, provided the steam and water circuits are properly drained and reasonable care is taken in sealing the idle unit according to standard recommendations of condenser and turbine manufacturers. Copper alloys are also susceptible to corrosion/erosion damage when circulating water contains air that is released at partial vacuum and higher temperature in the exit region or when turbulence occurs.

B K MITRA, *Philadelphia, Penn*

Check the twin for leaks, too

Assuming that the tubes here are rolled into the tubesheets in a conventional manner, there will often be a small step in the tube at the end of where the rolling device or expander deformed the metal. Amount of deformation depends on the fit of the tube in the hole and on amount of metal stretch by expanders. The step is the point at which stress-cracking originates. Aluminum brass tubes are subject to season cracking, because of their high zinc content, when residual stress is high and atmospheric corrosion can occur.

The failure mechanism appears to be ferrous attack on alloy grain boundaries. Concerning the severe corrosion at the inner surface of the tubes, all the above conditions apply, plus aggravation by the corrosive attack. This gave the radial cracking an initiating point from which it could continue into the already stressed brass. I suspect that there are tube failures in the operating condenser, in the same locations as in the shutdown one.

B B BROWN, *Hagerstown, Md*

Look at sediments for clues before drying

Analysis of the sediment in the river water might reveal a potentially corrosive material not found in the condenser cooling-water analysis. Make the sediment analysis on a fresh wet sample taken at the same time of year as the beginning of the 16-month shutdown. Even better would be an analysis of the wet slime and sediment on the bore of the tubes where the cracked tubes were. My observation is that often the first part of a shutdown can be troublesome, giving far more corrosion than during actual on-line time. A long wet shutdown can worsen the situation, however. Therefore, on shutdown, thoroughly clean the waterboxes. Some operators go further, drying the tubes with warm air to remove water traces and kill residual slime. Once dried, the condenser should be kept dry. Finally, slime control, by chlorination, say, should be considered for the period preceding shutdown.

C L Bulow, *Huntington, Conn*

Nitrogen blanket gives protection from oxygen

Some suggestions that IGK may find helpful:

■ Stress-corrosion cracking requires simultaneous presence of tensile stress, corrosive environment, and suitable metallurgy. Perhaps overly rapid shutdown or startup induced thermal stress. Also, during shutdown, oxygen may have leaked into the tubes. Micropitting induced by oxygen could enhance crack propagation for stress-corrosion cracking. A check of shutdown and startup records would be advisable.

■ For prolonged shutdown, IGK could consider a nitrogen blanket to keep away oxygen. A less desirable alternate could be flooding with demineralized water containing hydrazine.

■ Recent literature on corrosion suggests that the mechanism of crack propagation is by breaking the protective oxide film. Zinc levels above 15% in nonferrous alloys make the metal prone to stress-corrosion-cracking attack.

G C Shah, *Houston, Tex*

Retubing is long-term answer to ammonia attack

Ammonia attack on the cooling-water side, followed by stress-corrosion cracking, caused the failure. In the 16-month shutdown, there was plenty of oxygen available to attack the protective oxide film. Biological action of organisms on nitrogenous matter, especially in once-through surface water, can produce ammonia. Since approximately 8% of the condenser tubes are out of service, and the chemical damage has been done, it's only a matter of time before minimum tube surface area is reached, calling for condenser retubing. The long-term solution to the problem is the complete replacement of tube surface with a less susceptible material, such as 304 stainless steel or titanium.

R J Smogor, Jr, *Homer City, Penn*

Ammonia stress-corrosion cracking is danger

Ammonia stress-corrosion cracking, caused by a combination of ammonia, oxygen, water vapor, and stress, is a possible cause of IGK's failure. With admiralty tubes, this problem is one that occurs during out-of-service periods, because the oxygen needed for failure would be available only then. In addition, ammonia-vapor concentration would be less when on-line. In a similar situation at Coffeen Unit 1, failure was pinned on a relatively small amount of ammonia solution leaking into the condenser during an outage. Operator error and condensate-recirc valve leakage combined to allow leakage, which occurred during a 20-week major boiler/turbine overhaul. To prevent recurrence, stringent ammonia pump valving procedures are in effect now for all outages.

J C PRESTON, *Decatur, Ill*

Reduce biological activity with treatment program

This reminded me of the problem at Powerton, where the once-through condenser developed 2600 leaks out of 28,000 tubes after a three-month outage. The leaking tubes had concentric nodules concentrated on the bottom inside, along the first three feet. Corrosion had removed the protective oxide on the stainless steel tubes. Decision was to coat the beginning three and one-half feet of the remaining tubes with 0-60-mil-thick epoxy, then soak the condenser with a solution of methylene bis-thiocyanate to destroy remaining bacteria. To cut further risk, the station decided to inject hypochlorite into the circ water for 30 minutes immediately before an outage, to reduce biological activity. Tube drying during an outage was also going to be looked at, along with an attempt to run with maximum circ-water velocity to scrub the tubes.

W D POULTER, *Cohasset, Minn*

1.4 Ways to wash out plugged heater tubes

The tubular air heaters on the back of our pulverized-coal, 650-psig, 710F boilers have given trouble from plugging within 3 ft of the top outlet tube sheet. The pattern of tubes and fouling is as sketched for two heaters, each with 1330 2-in-O.D., 14-gage tubes, 34 ft long. A third heater has 1248 2½-in-O.D., 14-gage tubes, 44 ft long, on slightly larger tube centers. These tubes also have corroded. All have been replaced after about 15-yr life. Flue gas is 650F in, 315F out.

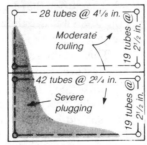

1.4

We have been air-turbining the tubes every 6-12 months, from the 4½-ft-high plenum space above the top tube sheet. Each plugged tube requires 15-40 minutes of cleaning time, and the method is slow and rough on our people. We understand that washing may be better.

What water pressure and flow rates do POWER readers use in this work? What kind of nozzle and what heater arrangement would be best? What is time required for each tube? And are there any other ways to clean air-heater tubes besides turbining and washing?—SG

Water is superior to steam or air for cleaning

We have never cleaned air-preheater tubes by the methods that SG asks about, but we have compared steam, air, and water jets on many types of deposits. The result was usually that the high-pressure water did a superior job for less money. Steam and air are best for sootblowing, where the deposits are hot and new, but where an encrustation has been building for a long time, the water will do a superior job.

Warm water could be necessary here, but it need not be condensate. As long as the waste can drain out an adequate opening at heater bottom, there will be little danger of corrosion. Adding chemicals to any water jetted in might cause heavy corrosion down along the present tubes in the region where they are now free from corrosion.

R M MALLON, *Kansas City, Kan*

Water experiments are first step in washing

Tubular air preheaters have always been troublesome components, and SG is not alone in his problems. We will be seeing more of this as we return to coal, too, for industrial steam and power generation. Turbining and washing are the only two methods that I have ever encountered for cleaning fouled tubes in this equipment. Neither method is as fast as operators would like, either.

The times of 15-40 minutes per tube seem a little slow, especially if all the plugging is within 3 ft of the top tube sheet. The indication is that the deposits are highly resistant and adherent, unless there is something wrong with the air-turbining technique. This I can't believe, because air turbining is such a standard procedure now. The only conclusion that I can draw is that the deposits are indeed tightly bonded to the tubes and are themselves hard.

If high-pressure water is to be the choice, SG will need to put on plenty of pressure to dislodge his deposits. This means getting a high-pressure jet unit, and experimenting with nozzles. Pressures will be in the region of 4000-6000 psig. Capacity could be as high as 50 gpm. The nozzle should be balanced, with at least three openings at 120-deg spacing, to make the lance easier to handle. A nozzle opening of $\frac{1}{8}$ in. should be good for the first trial. Slant the openings in the forward direction, too, so that the water will tend to penetrate obliquely under the deposits and wash them down as they are removed. The angle of the openings can be about 15 deg forward of perpendicular to the tube axis.

Tube damage from high-pressure water is possible, especially if the tube wall has been corroded partially through. SG should remember, however, that in most cases it will not be the water but rather the chemical action of the deposits and the acid in the flue gas that have weakened the wall. The experimental period is the worst time with hydraulic jets. The main point is not to become discouraged prematurely. Often a minor change is pressure or flow will improve the results considerably. The experimenting may have been done already for cases similar to this one. There are firms that specialize in high-pressure water cleaning of exchangers and other equipment. Perhaps at least one of these firms has had success in removing deposits from air heaters.

Significant things to look for when seeking help from an outside cleaning firm will be the exact nature of the problems it has worked on, in respect to the fuels and temperatures, and the makeup of the deposits, in both chemical and physical sense. If the factors are close to what SG has, then the experience of the firm can be relevant. If not, then the firm will be operating on past judgment of how to proceed on a fairly new kind of work. The success achieved will probably be little better than what SG could do by himself.

I tend to think that the first hydraulic-removal tests should be done without any preliminary wash water applied. The deposit will then be fairly dry, and there will be no physical or chemical action to tighten the hold on the tube wall. Start the jet at a point down the tube, at the bottom of the major deposit zone, and slowly pull the pipe upward. After doing a few tubes, check on the results and keep track of what has happened. After a change or two in pressure and nozzle opening, SG should be able to tell whether the water-jet method is better than turbining.

F D JOHANSEN, *Dayton, Ohio*

Higher exit temperature will help

The 305F temperature seems a little too close to the safe lower limit where SO_2 is concerned. If SG were to run at a slightly higher exit temperature, say 325F, in my opinion he would have less trouble from deposits and corrosion. Secondly, the pronounced tendency for deposits in one corner indicates that gas flow is not as uniform as it should be. Perhaps a baffle in the outlet plenum could help with this, so that the deposit problem would not occur so quickly.

The life of 15 years for the tubes is not bad at all, however. Average life such as this, plus the constant recurrence of deposits, indicate that the problem is not so much from a chemical corrosion effect as from physical causes that are leading to agglomeration and flyash settlement in a sludgelike condition. Besides raising the temperature of the exit gas, SG could ask several of the water- and fuel-treatment chemical manufacturers about fuel additives that might reduce the deposit tendency.

R U MARASCO, *Los Angeles, Calif*

Choose warm boiler feed at 200-300 psig

Water washing air preheater tubes of coal-fired boilers has been satisfactory, provided certain procedures are followed. Warm boiler feedwater is preferable, because it has only limited quantities of acid-solution forming elements.

This advantage is especially important if the heater will not drain completely. Pockets in parts of the heater-tube circuits can hold liquid.

An alkaline rinse, followed by drying out, will hold down any acid-corrosion problems. Do not wash with alkaline solution before the initial wash with boiler feedwater, however, because the alkaline solution will react with acid-corrosion products. The result will be that nearly insoluble precipitates will adhere to the tubes and cause increased plugging.

Compressed air, steam, and water have all successfully cleaned preheater tubes. Air has the advantage of not needing condensate drainage. In water cleaning, it is easy to vary pressure and flow but, as with steam cleaning, adequate drainage is needed. Pressures, flow rates, and cleaning times depend on available net flow area inside the tubes and on bond strength between deposit and tube wall. Typical pressure values are:

- Air, 75-200 psig
- Steam, 90-250 psig
- Water, 200-300 psig

Fluid pressure must be low to prevent erosion.

Contributing to the problem of deposits in air heaters is low-load operation, usually accompanied by reduced gas temperatures. The resulting corrosion of heater tubes frequently occurs at the cold-air inlet end. If there is higher-than-usual draft loss and/or higher-than-usual exit gas temperature, plugging is indicated. Monitoring these two variables will allow SG to decide when to clean the unit, in adequate time before fouling gets to be excessive.

If the draft loss and exit gas temperature become too high, then a partial or complete gas bypass around the heater, or a recirculation of hot air to mix it with the entering cold air, will usually retard the harmful effects until shutdown and cleaning.   H B WAYNE, *Jamaica, NY*

1.5   Operating 150-psig boiler at 15 psig

We operated an oil-fired D-type boiler for several years before retiring it temporarily. The boiler is rated at 75,000 lb/hr and 150 psig. We need part of the steam capacity now for heating, but would like to drop down to a lower pressure to be on the safe side with this piece of equipment, which has some tube corrosion. We wanted to go to 40 psig at first, but we heard many adverse comments on that. The manufacturer advised against it. He said that we would change the circulation and heat distribution in the boiler and that we could run into trouble from tube overheating and scaling. Inspectors, too, have advised us against the reduced-pressure conversion.

Another idea is to go down to 15 psig, where we will no longer have to be concerned with inspection. We think that our steam lines can handle the larger volume flow. We don't want to buy new boiler capacity now for the heating load if we can use this old boiler. Can POWER readers tell us what problems we will face with the 150-psig boiler operating at 15 psig? Could we assist circulation by a pump, and will we need to change water treatment? What capacity can we reasonably expect?—AEB

Steam-density change is important at low pressures

AEB must have a natural-circulation watertube boiler. These units have a well-defined water-circulation path within the many parallel circuits of the tubes. Driving force for the internal circulation is the relative difference in water densities between warmer and cooler sections of the boiler. This density difference, however, tends to work against circulation at low pressures. Steam volume becomes so large that it may choke flow inside the tubes. If severe enough, the tubes overheat and fail. Never operate a boiler at pressure below the manufacturer-recommended minimum. Even at 15 psig, a sudden tube failure would be serious.

With the manufacturer's agreement, AEB may be able to operate at a lower pressure—if he is willing to derate the unit. Limiting heat input limits steam volume generated in the tubes, so that circulation can be at an acceptable level. Resize safety valves and other support equipment, however. A pump to assist circulation, or an addition of external downcomers, would be impractical on an existing setup. Instead, AEB should have the manufacturer or boiler inspector survey the unit to ascertain what is necessary to return the unit to full service. Restoration would probably be cheaper than the other alternatives.

R A BLYTHE, *East Stroudsburg, Penn*

Derating gave this plant trouble with fouling

Derating of a boiler like AEB's is fraught with problems—as we soon discovered after derating a 50,000-lb/hr unit. With the same furnace volume and burner-management system, a modulating or start-stop pattern develops. The temperature excursions tend to accelerate tube scaling and corrosion, especially with reduced circulation. Operating at 150 psig, with a turbine as reducing valve, seems the most practical and economic solution. The turbine could be the prime mover for a fan, pump, or compressor, reducing plant electrical load.

A A CLARKE, *Jackson Heights, NY*

Safety valves need much larger opening area

If the operating pressure of a D-type boiler is lowered drastically, the safety valves need a much larger opening area because of the larger volume of steam generated. Also, the boiler will be more likely to prime. If present, a superheater may give trouble also in a lowered-pressure situation, because the steam may not divide equally among the elements.

If the boiler has to be operated at a lower Btu/hr load, the lower furnace temperature may cause trouble in an oil-fired setup. One solution: Increase the refractory surface in the furnace. The better solution, however, might be to operate at 100 psig and make the 15-psig steam through a reducing valve or heat exchanger.

C A SHUTTLEWORTH, *Bethlehem, Conn*

Recommends going to hot-water heat instead

AEB should base his decision on these factors:

Heating-load requirements. It is never economical to operate a boiler below 50% load. If the boiler is too large, sell it and install a smaller unit.

Hydrostatic test. If the heating load exceeds 38,000 lb/hr, verify boiler condition by hydrostatic test. A boiler unsafe at 150 psig is just as unsafe at 15 psig. The only difference between a "high-pressure" boiler and a "low-pressure" boiler is the safety-valve setting.

Inspection. Changing to 15 psig will not eliminate the need for inspection. If the state does not require it, the insurance carrier will. Remember that only an atom bomb is more powerful than a steam explosion.

Hot-water heating. If the boiler is found or made reasonably sound, the best choice is heating with hot water by a heat-exchanger system. This allows steam generation in the 100-125 psig range. With all condensate going back to the boiler, little makeup and water treatment is necesary, and the unit should give many more years of satisfactory service.

M A SHELTON, *Portland, Ore*

Boiler was not designed for heating mode

Operating an existing 150-psig boiler at 15 psig will eliminate the need to meet the requirements of ASME Boiler Code Section I. The boiler, acting as a heating boiler, should then meet Boiler Code Section-IV requirements. State and city may have adopted these heating-boiler code requirements, or they may have their own special requirements. In most cases, if the boiler meets Section-I requirements, it will also meet those of Section IV. The inspector will check the boiler and recommend tests if necessary to recertify the boiler.

The heating surface, heat distribution, and circulation system of this boiler are not designed to operate economically in the heating-boiler mode, however. Options are to repair the unit to an acceptable standard, or sell it and buy the needed size of heating boiler. A boiler with corrosion must be closely watched, too. A hydrostatic test is one way to check tubes for leaks and strength.

N DESAI, *Livonia, Mich*

Is low-pressure steam available now?

Low-pressure steam is good for only a limited number of applications, such as space heating or steam tracing. Most plants do have excess low-pressure steam, so AEB should make a steam balance to determine overall economics. Boilers operate on a natural-circulation thermosyphon principle; circulation rate will obviously differ at 15 psig. A bigger fraction of tube length will get sensible heating. A computer solution is necessary for precise determination of circulation pattern and heat transfer.

For reliable operation, startup precautions should include chemical treatment (repassivation). AEB should find out if the corrosion is external or internal. No overall statement is possible, because, for one reason, a poor deaerator will give pitting and $Fe(OH)_3$ scale. A deaerator check would include steam distribution and quantity needs and chemical-treatment requirements (sodium sulfite or hydrazine). Reduced steam pressure will call for more deaeration steam.

G C Shah, *Houston, Tex*

Check for meter-caused pressure drop

Comparison of enthalpy and specific-volume data allows the conclusion that, with a constant feedwater temperature, the unit can produce about the same amount of steam per hour at either pressure. Ability to take sudden load swings will be lower at 15 psig than at 150 psig, however, which has an influence on the automatic-control system. Heat reserve in the boiler at 15 psig will be about 30% less. Slowing down the combustion control and letting the pressure drift more is a suggestion, if the system is to operate at 15 psig.

Meter trouble may occur in steam off take line. In an 8-in. line, orifice Δp would be about 140-in. H_2O with permanent loss of 70 in. or 2.5 psi. At 15 psig and the same 75,000 lb/hr, the orifice Δp would be about 560-in. H_2O, and the permanent loss about 280 in. or 10 psi, which is impractical. Increase in line size, a flow nozzle instead of a plate, or operation at reduced load are steps that may be needed. A quick estimate indicates down-rating to about 40,000 lb/hr, if pressure drops are to remain the same in the system. Control-system changes, and changes in steam-flow and perhaps air-flow metering may be in order, too.

Water treatment needs no change. Consulting the boiler manufacturer on whether a pump will improve circulation is recommended. If circulation is stagnant and a tube becomes steam bound, local hot spots may cause localized corrosion attack and then tube failure.

A Watson, *Chicago, Ill*

1.6 Why can't brackets be welded to boiler?

Just over a year ago, we installed a new watertube boiler as part of an expansion at one of our plants. Work went along on schedule, and we hydrotested the boiler under supervision of the inspector. Right after the hydrotest, however, and before boiler startup, we discovered that we needed some brackets and small channels, to carry a walkway and support some small-diameter piping.

It would have been simple and convenient for the contractor to weld several 4-in. angle brackets and two 6-in. channels to pressure parts of the boiler, but our inspector told us that, once the hydrotest is passed, no welding is allowed on pressure parts. The welds that our contractor was suggesting were very small fillet welds, which he proposed to examine by dye penetrant as an extra assurance against cracks.

I understand that, in nuclear plants, some welding of minor parts to pressure-containing elements is permitted after hydrotest. If the weld is examined and found sound, I don't see why small brackets cannot be attached by welding. If that welding were to be the cause of a fatigue failure, a hydrotest couldn't detect the danger anyway. Have POWER readers had experience with this question? Is it subject to interpretation, or should the Boiler Code be changed to permit this kind of post-hydrotest welding?—HW

Thermal expansion is one danger from welding

Maybe HW was lucky that the boiler inspector didn't let him weld that walkway to his boiler. Attachments shouldn't be welded to boiler pressure parts unless it is determined that any additional stresses caused by the attachments won't exceed the allowable stress of the pressure parts. This is discussed in the Boiler Code.

The biggest problem with the walkway would be differential expansion. For example, pressure parts of a 400-psig boiler would be at about 450F, while the rest of the walkway could be at room temperature, perhaps 80F. Steel expands approximately 0.01 in. per foot for each 100-deg-F temperature change, or 0.037 in. in the example. If the walkway were 10 ft long, the differential expansion would be $\frac{3}{8}$ in., easily enough to cause a tube leak adjacent to an attachment weld. Of course, this wouldn't show up in the hydrotest either, but HW should at least be aware of the potential problem.

D C BURGESON, *White Bear Lake, Minn*

Take stress-relief need into account before proceeding

The ASME Boiler Code states in Section C4.410 that no welding on pressure parts is permitted except in accordance with advice from the authorized boiler inspector. Section C4.420 requires approval from the authorized boiler inspector before welding is undertaken on a pressure part in a boiler. The inspector may require a written procedure before the work begins, and he will demand that the work be done by a welder qualified in accordance with Section IX of the Code.

Section C covers repairs to boilers, as well as other operational functions. If it were to be argued that the additions proposed by HW are construction, then the previous Code sections dealing with boiler construction must apply. According to them, no welding is permitted after hydrotesting. The procedure suggested by HW, in which dye-penetrant testing is done, does not cover the very necessary stress-relief function. High stresses can result locally from small welds, and the stresses could lead to pressure-part failure.

While it appears to be a minor job to install a few small brackets, in reality they could lead to serious boiler damage unless designed by a competent engineer. HW's point about fatigue failure is well taken, and he is right about the hydrotest not disclosing a potential problem. In my opinion, the Code implies extreme caution in welds on pressure parts, and the inspectors use caution when they restrict work such as HW wants to do.

B B BROWN, *Hagerstown, Md*

The inspector is the key man after testing

According to the National Board of Boiler & Pressure Vessel Inspectors, addition to pressure parts of such welded items as brackets, ladder clips, tray-support rings, and studs for insulation or refractory lining which may be subject to inspection under the code, are considered as repairs. In these cases, the unit has been accepted by the inspector after witnessing the hydrostatic test, and that is the reason why the additions are considered repairs.

Repairs to boilers are to be made by repair organizations having one of these types of authorization:

■ National Board authorization—"R" stamp.

■ Authorization possessed by an organization in possession of a valid ASME Certificate of Authorization.

■ Jurisdiction authorization.

The organization responsible for repair shall prepare a "report of welded repairs" and submit it to the inspector for his acceptance. The inspector may require a pressure test (hydrostatic test) after the completion of a repair on a boiler or pressure vessel. For repairs of a routine nature, the organization may not be required to file a record of welded repairs, but the inspector may require a pressure test.

Before acceptance of a repair, the inspector must satisfy himself that the welding was done in accordance with R-302 (NBI Code, Chapter III), must witness any pressure test that he requires, and must assure himself that the other functions he deems necessary for compliance with the NBI Code requirements have been performed.

Therefore, everything is left to the inspector. He will decide whether he would like to rehydrotest the unit or accept the quality of welding and consider that it will not affect the pressure parts. The inspector must assure himself that there is compliance with the NBI Code.

N DESAI, *Toledo, Ohio*

Nuclear permission? Very limited according to code

The work proposed by HW is more involved than merely welding minor attachments and giving a dye-penetrant test in place of a hydrostatic test. HW must take into account:

■ The effect of attachments in producing thermal stresses, stress concentrations, and restraints on the pressure-retaining members.

■ The effect of welding in producing interior macroscopic flaws. Liquid penetrant is fair to good for detecting surface indications. Magnetic-particle and ultrasonic testing are excellent for surface and subsurface discontinuities but are no substitute for a hydrostatic test.

The nuclear code does permit attachments to be welded after hydrostatic testing, but only on piping, and with limitations. It should be pointed out, too, that this nuclear piping will be given a pressure test and nondestructive examination before plant startup and during its lifetime, in accordance with Section XI of the ASME Boiler Code.

HW should remember, too, that the manufacturer has the responsibility of assuring that the effects produced by the attachments are considered, and that the welding and all other requirements of the Code are met. The hydrostatic test applied and witnessed by the inspector will assure the safety of the pressure-retaining structure after fabrication or repair.

A J SPENCER SR, *N Attleboro, Mass*

A second opinion can sometimes help

I'm for the Code because I'm for safety first. The welding that HW contemplates seems insignificant, but it is classed as a repair requiring prior inspector approval. Welding is extensively covered in the Code and, in limited instances, welds of certain sizes and functions are permitted in work of the kind suggested by HW. In welding Code elements, a good rule is to seek the advice and approval of the inspector first. The inspector must assure himself and the jurisdictional authorities that the work, the welding procedure, the material, and the welder himself are acceptable under current regulations.

An inspector is not infallible, and another opinion might be a good idea, preferably obtained from the supervising inspector for the district or insurance firm of the field inspector involved. The Code maintains the pressure integrity of vessels through enforcement practices and close scrutiny of all subsequent field welding, but flexibility is provided by the National Board practice of accepting documented inquiries when differences arise in Code use or interpretation. HW might try to suspend the walkway from the building structure or erect it from the floor, eliminating the need for Code welding. (Space must be available, of course.) Remember when using these methods, however, that some localities have ordinances governing height, width, and proximity.

J E HILSON JR, *Scranton, Penn*

1.7 Finding all condenser-tube leaks

Our nuclear station has a problem that must occur often in any plant with condensers. The problem is potentially more expensive for us than for other types of plants, however, because of our high downtime costs. What sets the stage for this particular problem is a raw-water leak in our main turbine condenser. Upon suspicion of leak, we take out of service the part of the condenser near the suspected leak. Then we locate the leaking tube and plug it.

We then face a critical decision. Should we assume that this tube was the only one leaking, and go back on line as soon as possible? Or should we begin a painstaking search for other leaks, some of which may be very small? The cost of unscheduled output loss is very high for us, so we usually close up and get into full operation again if no other leaks are obvious. We have little to go on besides intuition at the time of decision. Sometimes we have been correct—but occasionally we have been quite wrong. We have good capabilities in chemistry and instrumentation. Can POWER readers tell us of any proven, quick method of finding minor leaks before putting a repaired condenser back on line?—FKJ

Freon method has saved time at nuclear station

Here is a possible method, which has reduced downtime by two to three hours in one nuclear station. A propane torch, a Freon-detection torch with side aspirator, can be set up on the air-ejector discharge. A lab technician can be stationed there to guard against fire and to watch for any trace of Freon "green," which would indicate Freon in the flame. Then, with all but one of the condenser manways closed, another technician can introduce Freon into the raw-water box. If leaking tubes are still present, the Freon will be drawn into the hotwell and appear in the air-ejector exhaust. Setting up and running the test will take only 30-45 minutes. If the torch flame shows no trace of Freon, the condenser can be closed. The plant should be started back up 2-3 hours faster than with other methods.

M C ABRAMS, *Covington, Ky*

Check condenser bottom for tube-run leaks

The leak in the tubes can be at the tube end or along the tube length. A leak along the length, which is rare, can be detected by monitoring the chloride concentration (conductivity) of the condensate in each section at the condenser bottom, if it is compartmented by ribs. Once a leak has been detected, FKJ can locate it by several methods, with his choice depending on turbine/condenser configuration and ease of making the test.

The flame test is old, but reliable. The flame of a burning candle held near the suspected leak area will be drawn toward the leaking tube when vacuum is being pulled on the condenser. A portable ultrasonic leak detector has been developed, too, which can pinpoint leaks up to 30 meters away. Plastic sheeting, fluorescin dye, and halocarbon are other methods.

R K JAIN, *New Delhi, India*

Vacuum drop shows leaks quickly

The vacuum-drop test can give a quick check for tightness in turbine condensers. Observe the loss of vacuum over a period of time when the air ejectors are secured. This will quickly give a measure of condenser tightness. Vacuum drop in a condenser with no leaks and in one with a small leak can look like the sketch (1.7A).

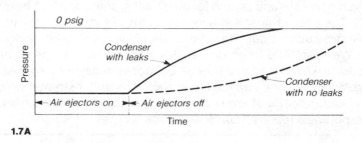

1.7A

Depending on size and age, a condenser with no leaks will normally maintain vacuum for several minutes to several hours after the air ejectors are secured. A leaking condenser will lose vacuum in just a few seconds or minutes, usually less than five minutes. Run this test at frequent intervals, say every three months, to find the no-leak curve for the base condition.

If leaks are present, the pressurized-water method is effective in finding them. First, drain and ventilate the circulating-water side of the condenser. Then clean and dry the waterbox, tubes and tubesheets. Compressed air will help. Next, fill the steam side of the condenser with feedwater, but don't flood the turbine. Inspection of tube ends on the tubesheet outside will reveal gross leaks.

Slight air pressure on the steam side, to 10 psig, helps to find small leaks. Be sure to test with warm water here, to prevent condensation, which will reduce sensitivity. Add various dyes to the water to improve sensitivity, but make sure that the dyes are chloride-free. Also, check with the condenser manual for design limits.

A Saran-wrap method requires draining of the circulating-water side of the condenser. Again, clean the waterboxes, tubes, and tubesheets, and dry them by compressed air. Apply a light coat of grease to each outside tubesheet face, and then apply Saran wrap on the face. Press the material around the holes to seal the sheet. Then draw the maximum possible vacuum on the steam side, but consult the manual for limitations. Observe puckering or possible failure of the plastic sheet over a defective tube.

We have used these methods frequently in checking tightness of turbine condensers. All the methods rely on already-installed instrumentation, so no special instrumentation is necessary.

M B TenEyck, *McKeesport, Penn*

Close off waterboxes before Freon

Try this method:

■ With water side of the condenser drained, close the inlet and outlet valves at the condenser waterboxes.

■ Charge the water side of the condenser with air containing a small amount of Freon.

■ Start up the air-removal equipment to establish a slight vacuum on the steam side of the condenser.

■ Monitor the exhaust from the air-removal equipment with a halogen leak detector. If leaking tubes are present, the detector will indicate Freon (see sketch, 1.7B).

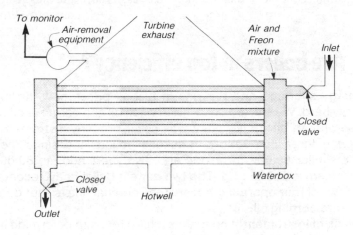

1.7B

By slowly raising the water level in the waterboxes, the approximate elevation of the leaking tube can be determined. When the presence of Freon disappears from the exhaust of the air-removal equipment, the leaking tube has become submerged.

E L CHAMPION, *Blackwell, Tex*

Pressurize steam or water side in this fashion

To test by vacuum on the steam side of the condenser, first remove the manhole plates on the waterboxes after draining the boxes. Then test each tube end on both heads with smoke tapers. Follow with a soapsuds test. Be sure to check the tubes row by row.

To test by pressure on the steam side of the condenser, remove the manhole plates on the waterboxes after draining. Jack up or block the condenser support springs or bearings. Fill the condenser slowly on the steam side with water and a fluorescent dye. Check each row of tubes as the condenser fills. Do not pressurize the condenser with more than 10 psig.

To check with pressure on the condenser waterboxes and tubes, dump the condenser, and have inspectors enter through the hotwell access plates. Fill the waterboxes and tubes slowly, checking visually and marking leaks on diagrams drawn on scuba slates. This is miserable work, but it can be done. Electronic testing methods for tube thickness and continuity are available, and can be part of plant equipment for scheduled downtime work.

J KING, *Brighton, Mass*

Dye and UV light make this task easy

Fill the steam side of the condenser with warm condensate and add a little dye that is sensitive to ultraviolet light. Then view the tube surface with an ultraviolet lamp. Tubes with minor leaks can be easily seen and plugged.

A P LITSKY, *Massillon, Ohio*

1.8 Are boilers at top efficiency?

The energy shortage and the rise in fuel-oil prices have hit my plant especially hard. We started up in 1962, when oil and gas were cheap, and because of that, our boiler plant was never properly instrumented. We began with three 50,000-lb/hr package boilers and since then have added a similar 60,000-lb/hr one and, two years ago, an 80,000-lb/hr unit. All steam is at 225 psig. The two newest boilers have economizers. Right, now, all four operating boilers (one original one is shut down temporarily) are burning oil.

Our lack of instrumentation makes it hard for us to determine if we are at optimum or peak efficiency. For example, although we have a steam-flow meter on each boiler, the meters lack integrators, and we have to planimeter the charts to get flow. Fuel-oil consumption rates are hard to learn because we don't have meters for each boiler—instead, we have been gaging the tanks.

I check efficiency by oxygen analyzer periodically during the day to monitor excess air, but I wish we could have more assurance that our boiler controls are giving us best results. Repeatability of results is questionable for at least one boiler. What advice can POWER readers who have had similar experiences give me? What is the minimum equipment I need to determine efficiency accurately? And are lengthy boiler tests a must for my boiler plant?—KI

Suggests present instrumentation is enough

Although by today's standards, KI's instrumentation is spartan, it is adequate for determining boiler efficiency through the "in-out" technique in ASME Performance Test Code 4.1. Compare net heat absorbed in the boiler with total heat introduced in the fuel. Boiler efficiency is $1 - [$ (fuel heat input $-$

boiler heat absorbed) ÷ fuel heat input]. KI should start by a check, repair, and recalibration of all present test instruments. Then more instrumentation is necessary to measure the following:

■ Boiler output conditions are most demanding. Measure steam pressure with deadweight testers calibrated and corrected for any water legs. For steam temperatures, thermocouples and an accurate bridge are musts.

■ Condensate temperatures are found by calibrated mercury-bulb thermometers. Condensate pressure is secondary in importance. A calibrated gage is good here. Condensate-flow recorders are useful to check the steam-flow recorders.

Gauging the oil tanks is acceptable during a boiler test if one tank can be isolated to feed only the boiler under test. With preliminaries completed, adjust loading of all boilers so that the boiler under test will operate steady for about four hours. During that time, fire the boiler to give normal O_2 levels characteristic of everyday operation. Record steam flow, steam and condensate pressures, and temperatures every 5-10 minutes. Gauge the oil-tank level at the beginning and end. Repeat the procedure for various boiler loadings, such as 25%, 50%, 75%, and full load. For data interpretation, find steam and condensate enthalpy for the average steady-load readings. Difference between steam enthalpy and condensate enthalpy, multiplied by steam flow, gives net heat absorbed. High heating value of the oil times total oil in the test gives the gross heat to the boiler. The formula above then serves to get efficiency.

This method is accurate and repeatable. Additional tests with O_2 levels varied will show the effect of improper firing on boiler efficiency. The effects of increased combustion-air temperature or better sootblower operation are also of interest. With all this information, KI would have a basis for operating the most efficient boilers at higher loads and for longer periods. This reduces the burden of the less-efficient units. The technique is called "incremental loading" by the utilities, to whom optimized efficiency is vital.

P J Russotto, *Staten Island, NY*

Get oil flowmeters first and analyze combustion

KI has an annual fuel cost of probably $5.5-million. In view of this, just about any combustion-control supplier would install the necessary instrumentation to control and monitor efficiency for the price of fuel reduction of one year. Any noncontrolled plant should be able to reduce fuel consumption 1% to 5% with proper controls. Integrators in KI's flowmeters should be a good low-cost first step. Boiler outlet temperature at any steam flow gives a good indication of excess air. Oil flowmeters for each boiler are a must. With these additions the operators can check steam/fuel ratios and spot-check at any load point over a 10-minute period on any boiler.

KI should get combustion analyzers, too. Change in oil viscosity or heat value will change the required air/fuel ratio and combustion efficiency. Analyzers will detect excess or insufficient combustion air and poor atomization. The latter is a subject in itself. Air-flow changes cannot correct for poor atomization. KI should concentrate efforts on the boilers with economizers, because their better steam-to-fuel ratio warrants better instrumentation.

G A SCATKO, *Tampa, Fla*

Figure in all boiler taps to find amount of steam

Our situation was similar to KI's, except that we had oil meters but no flow meter on each steam generator. The efficiency equation we use is:

$$\text{Efficiency} = W_s \times (h_g - h_{fw}) \div W_f H$$

where W_s is the steam rate per hour from the generator, h_g is the enthalpy of the steam, h_{fw} is the feedwater enthalpy, W_f is weight of fuel per hour, and H is the heat value of the fuel in Btu. A flowmeter is a must for W_f, and a thermometer for h_g. For W_f, an oil meter is also a must, and the fuel supplier will give a certified fuel analysis including H.

The hard part is next. For an accurate determining of boiler efficiency, KI must find the amount of steam in his continuous blowdown, sootblower operation, mud-drum discharge, and any other taps from the boiler not picked up by the flowmeter. If these are not added to the flowmeter reading, the efficiency reading will be false. For example, 1-2% of output can go to continuous blowdown. KI will also need means for condensing the flash or cooling the high-temperature water. Finally, on the combustion side, KI should have an orsat apparatus for finding CO and CO_2 to determine combustion efficiency. These are in addition to O_2 readings.

B B BROWN, *Hagerstown, Md*

Good operating practice beats long testing

Because average everyday boiler efficiency is usually well below the efficiency found under test conditions, lengthy testing wouldn't satisfy KI's needs. I would suggest that he use an orsat and also follow good operating practice, such as keeping the settings tight, the fuel burning-equipment clean, and the fire and water sides clear. Boiler blowdown should be the minimum recommended, too, and each unit should operate at its most efficient rating, except for the one taking load changes.

W C MCCARTHY, *Brant Rock, Mass*

Measure feedwater flow rate, temperature

Since steam-flow orifices have flow variations from steam pressure and quality differences, KI will find feedwater flow to be more accurate and practical. Each boiler unit should have a high-pressure hot-water meter to measure and record flow. Feedwater temperature must also be taken to get water density. Under stead-state load conditions, with no blowdown, the feedwater flow rate will equal the steam rate.

L L YOST, *Winfield, Kan*

Heat-loss method is best application here

Extensive boiler tests are unnecessary for the plant KI is operating. Although it would be nice to have a separate oil-flowmeter and steam-flowmeter with integrator for each boiler, KI can find boiler efficiency without them. The two methods in the ASME Performance Test Code 4.1 are in input/output method and the heat-loss method. Because KI lacks instrumentation, the heat-loss method is best.

Losses to be determined are from dry gas, moisture in fuel, H_2O from H_2 combustion, combustibles in ash, radiation, and unmeasured losses. Major item in these is the heat loss from dry gas. If excess air is at a minimum and the boiler exit-gas temperature is low, the heat loss to dry gas will be minimum, too. KI can install a dry-type oxygen probe and thermocouple to measure boiler-gas exit temperature. This can be monitored in each boiler to assure that operation is near optimum. In an orsat analysis, the sum of O_2 plus CO_2 is usually a constant of 13 to 17 for oil firing.

The excess-air value for oil firing averages about 10%. KI should keep stack temperature near the design value, since he loses 1% efficiency for every 40-deg-F increase.

R B DEAN, *Muscatine, Iowa*

1.9 Erosion and cracks in boiler drums

Our natural-gas-fired generating station in Pakistan has four identical 600,000-lb/hr single-drum units operating at 1400 psig, 960F. Two units (1 and 2) started up in 1960, and two (3 and 4) in 1963. In the first two years of operation on Units 1 and 2, very deep erosion and cracks developed at both dish ends of the steam drums. The rest of the drums has stayed free of any defects. We ground out the cracks and eroded areas at various locations around the manhole. The next 10 years were uneventful, but recent inspection of the drums showed more cracks at different locations and considerable erosion on areas not ground during initial repair. Drums of Units 3 and 4, on the other hand, are entirely smooth and show nothing wrong. Operating conditions and water treatment are exactly the same for all generators.

Drum dimensions are 82-in. inside diameter, 414 in. long, 4.33-in. wall thickness. Material is German. We do not know whether the steel of all drums received the same nondestructive testing, because we have been unable to get replies from the German manufacturer. This problem is like a millstone to us. Can POWER readers tell us if this is a common problem? What precautions should we take to protect our boiler drums from this damage? And why don't Units 3 and 4 have the same trouble? —PAK

Stress corrosion blamed for drum cracks

The problem appears to me a typical case of stress corrosion. In almost all cases, this originates from improper treatment of boiler water. Even though the chemical treatment is believed to be the same for all boilers, it would be wise to check again to make sure that the right amount of chemical is being fed to each boiler. A choked line, faulty pump-capacity setting or leaking valves or pump-head gaskets may be jeopardizing the value of what are thought to be identical systems. One good way to investigate the problem might be a regular check on pH of drum water for all boilers.

B K MITRA, *Philadelphia, Penn*

Compare water-treatment records day-by-day

Because troubles like this are not common in the US, it is difficult for anyone here to fathom the exact cause. Although metallurgical reasons may be right up front in explaining the mechanism, I think water treatment is a vital contributing factor. If the water treatment were correct and kept up to specification, the triggering action of unsuitable water would no longer act to set off attack on the susceptible metal of the drums in Units 1 and 2. I would consider the following plan:

■ Go over all the water-treatment records, right from the beginning, for all boilers. This means day-by-day study, not merely listing the first day of each change in treatment.

■ Check through the chemical-feed systems to verify delivery rates and to see whether shot or intermittent feeding could cause temporary highs in concentration or pH.

■ Compare in detail the records and feed systems for Units 1 and 2 against Units 3 and 4. This would include load on the boilers and number of times on standby. If one pair of boilers had been cycling more than the other, this might be a good clue.

■ Look for a difference in water circulation near the drum ends in Units 1 and 2 vs Units 3 and 4. Sometimes boilers that are "identical" but differ by a couple years in life have had design changes that seem minor but may be enough to make a big difference in corrosion performance of individual parts. Circulation in tubes is one example of this.

Cooperation of the boiler manufacturer is a must if PAK is to get at the metal analysis and history. Proper samples of the boiler metal would be hard to take now.

P T CONCANNON, *Chicago, Ill*

Is erosion the problem, or corrosion?

If the erosion and cracking are only on the drum ends, then the problem almost certainly involves the metallurgy of the heads. Water conditions in the drum may contribute to the problem, but they cannot be the only cause. The fact that Units 3 and 4 have been trouble-free for so long is further evidence for this theory. If the metal in the drums differs from the head metal, there might be attack at or near the circumferential welds. This should be checked, too.

The full extent of the defects should be looked into very carefully. Although both erosion and cracking start from the surface, there can be no assurance that subsurface defects are not present unless a thorough examination by radiography or ultrasonics proves their absence. Once this is done, attention can focus on the exact mechanism of attack.

What appears to be erosion may prove to be corrosion instead, with the products washed out into the moving water in the drum. This should be checked carefully by noting the exact location of the eroded areas with respect to the water line. Directionality and orientation of eroded areas could be important, too.

Another suggested step is to try to recover surface deposits from the eroded areas and their immediate vicinity. As a complement to this, metal and corrosion products removed in grinding the cracks should also be collected. All the corrosion products recovered, together with location information, should then go to a metallurgical consultant. It is astounding that PAK has not received replies from the manufacturer. A problem of this potential size should be getting every possible attention because of the inherent danger.

R W Horner, *Springfield, Mass*

Stop droplet condensation at dish ends

The erosion, if not the cracks or pitting, could be caused by erosive action of water droplets that are condensed at the cooler surface of dish ends and run down along the surface, as the sketch shows. An increased thickness of lagging might arrest this damaging effect. Also, if the units were operated at constant load, the damage from cracks might be reduced greatly.

N Nasrullah, *Multan, Pakistan*

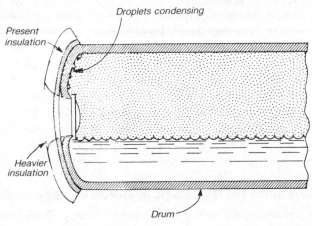

1.9

Galvanic action may play a part

The problem of boiler-drum erosion and cracks is not a common one. It is also quite dangerous for a boiler of this size. One possible reason is galvanic action of some kind. The material of the manhole cover may be different from that of the boiler drum, and, in the presence of chemically treated feedwater acting as an electrolyte, a local action may have been set up in the drum. During long shutdown periods when the boiler is heated periodically, this galvanic action between manhole cover and boiler-drum surface comparatively near the manhole could continue. A close visual inspection of the inside of the manhole cover might show some extra metallic coating that would confirm the possibility of this.

The original metal, too, could be responsible. Perhaps the erosion and cracks come from nonhomogeneity of constituents, and the wrong pH value of feedwater would then be a factor in the attack. Evidence of erosion and cracks near the manhole may support this.

J Atif, *Flushing, NY*

1.10 Stop cracking of burner refractory

Our oil-fired steam generators have been giving generally good service during the recent fuel shortage. The one real trouble we have had with them during the past year has been in the refractory work around the burners. We have experienced repeated cracking of refractory in the throat zone. We have been told that vanadium in the fuel oil could cause the cracks: Supposedly the vanadium compounds deposit on the refractory and then flex with temperature change at shutdown to start cracks. We were also told that in one boiler the burners were set back too far, causing conditions favorable to cracking. We hesitate to set the burners forward, however, for fear of disturbing flame pattern and impingement. In addition, piping changes would be necessary.

Because we had no trouble with burner refractory in the past, I think the fuel oil is responsible. As far as we know, burner locations and settings are the same as always, and loads and twice-yearly shutdowns are unchanged. Have Power readers noticed cracking similar to ours, and, if so, what are the remedies? Are there any new materials that we can try as a solution for our problem? Can we protect the refractory in any way and still leave our burner settings unchanged?—DWC

Make allowance for expansion, which causes cracking

Insufficient expansion allowance is the major cause of severe cracking. With prefired refractory tile, a burner throat will crack at the mortar joints when fired, and the cracks will become visible when cold. At operating temperature, however, the cracks will seal and give a tight, smooth surface. This type of crack is acceptable if the bricks stay in position. One kind of expansion joint is shown in the sketch.

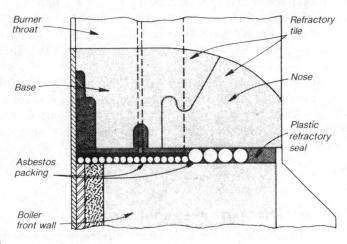

1.10

Oil spray from the burner must not impinge on the throat refractory at any point. If it does, then push the oil nozzle forward until the pattern clears refractory. Impingement, especially when vanadium is present, causes refractory to chip and spall and eventually erode away. With plastic refractory, venting and curing before starting is important. Otherwise entrapped moisture, unable to escape, will turn to steam and the refractory may explode instead of merely cracking.

L L YOST, *Winfield, Kan*

Try thermal dilatometric analysis

Perhaps DWC should put his refractory vendor and service department to work. The start would be an inquiry as to the refractory mineral composition changes and consequent thermal expansion characteristic changes. A return to the old formulation is suggested if there has been a change.

Another step would be analysis of the suspected vanadium-oxide deposits on the throat-area refractory and an attempt to duplicate the composition in the lab. Next, a thermal dilatometric analysis on it and on the refractory itself. The results will indicate any thermal mismatch, which could be the problem source. If mismatch does exist, the vendor should recommend an alternative refractory. Don't neglect the burner-heat expansion capabilities, however. Because DWC had no bad experience in the past, he should check his field installation procedure, too, to make sure that no changes have been made.

J M BARANSKI, *Jackson, Mich*

Plastic refractories serve well in England

The Central Electricity Generating Board developed a standard to overcome similar problems of cracking and spalling around oil-burner throats. Quality of refractory materials, method of making and securing the refractory to the furnace tubes, and initial prefiring curing period are important factors. A method of placing refractory is to divide the peripheral area into segments, using a material that will burn away quickly on light-off but remain in place for the curing period. Segments should be small enough to breathe with the tubes to which they are attached.

Currently we weld on split studs, Y- or V-shaped, instead of the straight ones. Our experience indicates no relationship between vanadium content of fuel oil and refractory cracking. Of course, DWC should be careful when steam cleaning or water washing combustion-chamber spaces, to prevent water from getting behind the refractory.

T R SHAW, *London, England*

Add cement slurry at every shutdown

In a similar circumstance, we mixed a slurry of refractory cement and painted it over the joints between the firebrick around the burner and the front wall. Every shutdown required renewal of the cement. To improve the situation, we would open the joint and insert a filler of asbestos wicking, then seal it with refractory cement. It all had to be replaced at shutdown. The factory representative said we were overfiring, but, as in DWC's situation, our No. 6 oil is high in vanadium.

B B BROWN, *Hagerstown, Md*

1.11 Why do tubes fail after shutdown?

In the past 15 years of operation of two of our 65-MW turbine-generator units, dezincification has forced us to retube the admiralty-brass tubes in the condensers three times. Very few defects occur during normal operation before shutdown. When we start up after a three-to-four-month shutdown for overhaul, however, we find hundreds of tubes leaking. This happens even though we dry and clean the tubes as far as possible. We clean the tubes with nylon brushes and water jets, then dry with dry air. Condenser doors and drains are kept open to avoid flooding from leaky inlet and outlet valves.

Our river water (Pakistan) contains 100-4000 ppm silt. The temperature in the turbine room remains over 100F—far above the dewpoint. Another baffling aspect of the problem is that most of the failures occur

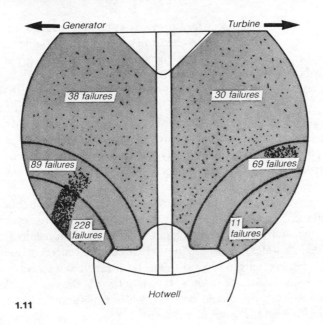

Generator ◄—   —► Turbine

38 failures   30 failures

89 failures   69 failures

228
failures   11
                     failures

Hotwell

1.11

in one area, with scattered failures elsewhere (see sketch). We considered the possibility of stray currents from the generator, but came to no definite conclusion. We have recently retubed with aluminum brass. Can POWER readers give us any idea of what causes our trouble? What is the best method to protect tubes from this type of corrosion?—NGP

NH₃ on steam side attacked lightly loaded tubes

We experienced a similar problem. We found out that ammonia concentrating on the steam side attacked tubes that had very little condensate collecting on them. This was in the air-cooler section, which was baffled off from the main drips.

T G MURPHY, *Westwego, La*

Try an eddy-current probe to observe corrosion

NGP raises an interesting problem concerning condenser-tube failure by dezincification. The details given bring out several issues for which further investigation might be useful. I have no hesitation, however, in saying that, if the trouble is actually dezincification, there is a very simple solution where admiralty brass is the metal—merely ensure that the tubes are appropriately inhibited. The most satisfactory inhibitor for this is a carefully controlled addition of 0.04% arsenic. Other materials, such as antimony or phosphorus, would also serve.

I suspect, too, that the condenser tubes do not, in fact, fail in service right after shutdown, but instead are found right after shutdown to have failed. In other words, the failure has occurred previously, but the leakage was prevented from becoming obvious until the tubes were cleaned and the debris removed. In some instances, perhaps the small plugs of dezincified material have actually been knocked out by the cleaning operation.

It seems to me most unlikely that stray currents influence the distribution of the failures. More likely, the two areas worst affected are the result of conditions on the outside of the tubes. It would be interesting to know if drains from auxiliary condensers, glands, etc, are connected to the condenser in these areas, and if the inlets are not properly screened or baffled. Dezincification would be encouraged if there were a local higher temperature such as might arise under these circumstances.

Another interesting observation program for NGP would be a tube probe in these and adjacent areas, using an eddy-current instrument to determine distribution of corrosion along the tube length. If the problem results from some external effect, the location would be indicated by occurrence of the most severely corroded portions in several tubes, both failed and sound, at approximately the same distance along the length of the trace. An eddy-current-probe examination could be worthwhile from another point of view, too: determination of the general extent of corrosion, and establishment of whether or not some tubes were completely free of dezincification attack. If some tubes had escaped attack, it might be because they contained an accidental addition of inhibitor. The dezincified tubes would almost certainly not contain inhibitor, or perhaps would have an impurity, such as magnesium, that would adversely affect them.

If the failure resulted from use of uninhibited admiralty tubes, and if the replaced aluminum brass tubes come from the same source and are uninhibited, then they will also be attacked by dezincification.

C BRECKON, *Methley, Leeds, England*

Failure may be from inside condenser tubes

Several reasons could be behind NGP's condenser-tube failures. Study of the surface at the failure point can give important clues, and here are some possible causes:

■ If the group of 228 failed tubes is near the air-cooler section, then ammonia may be responsible. It is readily soluble in condensate, with little subcooling. Ammonia concentration in an air-cooler section is 300-600 times that in the hotwell. Decay of organic material or excessive amine dosing could cause ammonia to be present. Stress-corrosion cracking from ammonia can occur during nonoperating periods. Grooving or smooth and uniform thinning is a sure clue to ammonia-bearing condensate.

■ Tube failure from inside may occur even though tubes are cleaned and dried as well as possible at shutdown. Tubes can still be dirty near the center. Then ammonia can form from decomposition of organic matter during a shutdown. This would explain the failure during shutdown, but not the local failure at the bottom of the condenser toward the generator side. This might result from erosion during normal operation, as a result of entrance of drips and drains here. In such a failure, the tubes will be found badly pitted from outside. Failure from inside can be avoided by cleaning tube insides with hydrochloric acid. In acid cleaning, several precautions are necessary. The steam side should be kept full of water, and venting of such gases as hydrogen and CO_2, formed by acid action on metal and calcium carbonate, must be done.

■ Erosion on the steam side, another possibility, depends mainly on flow distribution, steam velocity, and steam quality. It is possible that NGP's tube spacing in the worst-affected region causes high-velocity jets. This erosion, combined with ammonia corrosion, may cause the failures.

R K JAIN, *New Delhi, India*

100% humidity can promote corrosion

Dezincification is the result of local electrolysis: A local electric cell is formed by the metals of the tube alloy and an electrolyte, and the negative zinc is "wasted away." In NGP's case, I wonder if the tubes in question have been sectioned to indicate on which side (steam or water) the corrosion starts. Also, what is the pH of the condenser cooling water? The sketch with the problem indicates that exhaust steam impinges almost directly from the turbine exhaust, maintaining high temperatures in the greatest failure area, where corrosion would be most active.

Are the condenser watersides flushed and wirebrushed immediately after being taken out of service? Is the condenser laid up perfectly dry or filled with treated water after cleaning? If dry, is the drying done with forced, heated air, and are all openings sealed to prevent breathing after cleaning? All this will prevent or discourage incipient corrosion cells. Once the cells have started, 100% humidity will promote corrosion even if ambient temperature is high.

Highly silted river water for circulation indicates high organic salt solution and an acid pH, which causes dezincification. When the condenser is flushed, a strong hot soda solution will neutralize acid deposits. Follow this by a hot clean flush of feedwater.

H M NEUHAUS, *Santa Monica, Calif*

Leak distribution is clue to type of corrosion

In investigating a condenser-corrosion problem, the following factors are important:

■ Whether the corrosion started from the water or the steam side.

■ The type of cooling-water treatment.

■ The type of feedwater treatment.

■ The location of the condenser's air-cooler zone, from which noncondensable gases are drawn by the ejector.

In NGP's case, however, the distribution of leaks provides a clue. The largest number of failures is in a relatively small area. If this area reasonably coin-

cides with the air-cooler zone, the trouble may be attributed to ammonia corrosion. Ammonia in one form or another serves to condition drum water. We
have replaced tubes in the air-cooler zone by Type 316 stainless steel to rectify
conditions like those of NGP.

B K MITRA, *Calcutta, India*

Check inside surface of tubesheet right away

I think NGP's condenser problem is caused by electrolytic corrosion. The fact
that the majority of corroded tubes are in the lower (or liquid) part of the condenser, rather than in the upper (or wet steam) part, seems to support this. The
remedy is simple: install a cathodic-protection system. NGP does not mention
the material of the tubesheet. It would be advisable for him to take a close look
at the sheet's inside surface, especially the region where the greater number of
damaged tubes connect. A bad surprise may be waiting for him. Another reason for dezincification could be high pH of the cooling water. Nevertheless,
this does not seem to apply here, because if it did, then all tubes should be corroded more or less evenly, which is not the case.

J S CRISI, *Rio de Janeiro, Brazil*

Holes from corrosion came before shutdown

Years ago, a local water company supplied domestic water to our area from
wells. The water was appreciably alkaline, and dezincified brass pipe. Leakage
was not immediate, however, because the zinc-alkali deposit had greater
volume than the displaced zinc. The "rock" deposit made the pipe outside look
bad, but the piping held water. When the wells were shut down and soft Croton
water sent through the mains, however, the zinc-alkali deposit dissolved, and
leakage resulted.

Perhaps NGP's trouble has a similar cause. His water must be similarly alkaline, and the deposits plug the dezincification capillaries, holding tightly as
long as the deposits are kept wet. During shutdowns, however, the zinc-alkali
deposit loses its water of crystallization, turns to powder, and loses its plugging
properties. Therefore, it is flushed out of the interstices once the condenser is
returned to service. Then, of course, the tubes leak like a sieve. We found red-
brass or solid-copper tubing most effective for the alkaline water. In any case,
a water analysis is advisable, followed by selection of proper tubing materials
to cope with the cooling water. A M PALMER, *Brooklyn, NY*

1.12 Adapting pump to existing boilers

Our proposed new boiler-feed pump has brought up some questions.
Our industrial plant generates 50,000 lb/hr of steam from a 400-psig
boiler and 20,000 lb/hr from a 200-psig boiler. The present six-stage
main boiler-feed pump delivers 245 gpm at 1250 ft total head. Feedwater

regulators reduce 630 psig discharge to 505 psig for one boiler and 260 psig for the other. The new pump has eight stages, 1440 ft total head, and 850 psig discharge. Motor is 150 hp. Test shows 70,000 lb/hr discharge at 775 psig.

Can POWER readers tell us if this pump can be run continuously at a system overpressure of 145 psig at normal load (and 220 psig at half load)? Can we replace two opposed-stage assemblies by two dummy stages to cut discharge pressure to 630 psig? Or would eight reduced-diameter impellers be a better solution? Would increasing bypass flow back to the deaerating feedwater heater solve the problem, and would it affect deaerating ability of the heater? Finally, is it practical to lower the pump speed by change to turbine drive, provided we find a use for the exhaust steam?—WR

Check effect on deaerator performance

Assuming that WR's 70,000 lb/hr is the maximum demand, the highest pump power at 60% efficiency would be 105 hp. A turbine would require this equivalent from the boilers and would not be recommended. Recirculation back to the deaerating heater is acceptable provided the return line entrains no air. The flow must enter below the water line and not disturb normal flow of oxygen bubbles to the surface. Operating the pump at 850 psig will not harm the system, but dummy stages are feasible to reduce the head. The feeling persists that someone sold WR's organization more pump and motor than was necessary. The 150-hp motor at two-thirds load will reduce his power factor, unless he installs corrective capacitors.

B B BROWN, *Hagerstown, Md*

New pump seems undersized for application

If WR can use exhaust steam, then a turbine drive should be installed to match delivery from the pumps to actual needs. Power obtained as byproduct from reduction of steam from turbine throttle to exhaust pressure will be cheaper than purchase of electricity on an out-of-pocket basis. Bypass of excess flow to the deaerating chamber may affect deaerating performance by flooding the trays or sprays and preventing proper exposure of new incoming water for release of gases. Because the bypassed feed has already been deaerated, however, it can safely be returned to the *storage* compartment of the deaerator.

Either of the above expedients is attractive, because either will allow the full capabilities of the new pump to be retained. In regard to steps that would alter the balancing provisions of the new pump, however, perhaps WR should refer the idea back to the manufacturer. Otherwise he might end up with an unbalanced pump. Reduced-diameter impellers would be preferable, but that question, too, should be taken up with the manufacturer.

The sizing of the two feed pumps is something to wonder about. WR's boilers total 70,000 lb/hr and, assuming 10% blowdown, they would call for about 165 gpm at maximum delivery and 505 psig. The old pump is about 50% high in capacity and close to pressure needs, but the new pump is undersized by the blowdown margin, and it is about 50% high in pressure delivery. A more customary allowance is about 25% to 35% in both capacity and pressure. This would call for a pump of about 210 gpm at 650 psig pressure.

M B GOLBER, *Chicago, Ill*

Take out two impellers

It is not clear to me why WR's company is planning to buy a new pump designed for 850-psig discharge pressure when the two boilers operate at 400 psig and 200 psig, respectively. I suspect that the ultimate plan is to replace the two existing boilers by a single boiler operating at some higher pressure, say 650 psig, and that the new pump is being selected for this. The possibility affects the relative merits of the various solutions, but we also need some assumptions:

■ Deaerator pressure = 20.8 psia or 6.1 psig
■ Temperature = 230F
■ Specific gravity of water = 0.952
■ Available NPSH = 20 ft or 9.1 psig
■ Total suction pressure = 6.1 + 9.1 = 15.2 psig

Converted to psig, the total head of 1250 ft is equivalent to 515 psig, so the pump would have a discharge pressure of 530.2 psig at 245 gpm. Assuming that the 630-psig discharge that WR mentions occurs at the 70,000-lb/hr flow to the two boilers (147 gpm), the total head is 1490 ft at the 147-gpm flow. A head-capacity curve can be constructed for the present pump.

No design capacity is given for the new pump, and the situation is less clear there—1440 ft total head corresponds to 595 psig, and it is difficult to imagine that the H-Q curve is so steep that at some part-capacity it develops the approximately 2020-ft total head needed to produce 850-psig discharge with a suction pressure of 15.2 psig. Nevertheless, WR has several valid solutions available:

■ Let the feed pump operate at overpressure. No particular harm would occur, but the regulating valves, throttling the excess, would wear a little faster and would waste considerable horsepower.

Remove two of the impellers. This is quite possible, but the two impellers should be opposing ones, as WR assumes. First stage should not be removed, however, or NPSH problems might result. Pump design will determine the exact configuration required for the two dummy stages. One thing is certain: That portion of the shaft left uncovered by impeller removal must be covered by dummy shaft sleeves.

■ Cut down all eight impellers. Because of the uncertainty of the H-Q curve of the new pump, I cannot calculate the exact amount of the cutdown. As much as 20% could be necessary. This would be close to the recommended limit for such a pump, assuming that the original impellers are near to the maximum diameter for this pump. If they are not, then the cutdown would be excessive, so WR would have to throttle even at full load of 70,000 lb/hr, all on the 400-psig boiler. This solution is not as efficient as the second solution. It is not the best solution if the pump will ultimately be used at or near its present design conditions, because the cut-down impellers would have to be replaced by full-diameter impellers.

■ Change to turbine drive. This is a very practical solution, particularly if the plant heat balance can utilize the exhaust steam. The variable-speed operation of a turbine drive would also give more efficient operation at part loads. Although a steam turbine is more expensive than an electric motor, especially on an installed-cost basis, the power saving at part load could offset the extra in capital expenditure.

Because the head developed by a centrifugal pump varies as the square of the peripheral speed of the impellers, it follows that it varies with the square of impeller diameter and/or with the square of running speed. Instead of cutting the impellers to 80% of present diameter, therefore, WR could reduce pump speed to 80% of present speed. The second and last solutions can be combined and the turbine drive will still save power when the speed is varied with load.

Increasing bypass flow to the deaerator would not affect the deaerating ability of the heater, because the bypass would go to the deaerator storage space. Nevertheless, this solution would be even more uneconomical than the first solution, in which excess pressure is throttled at whatever total flow is required by the boilers. With a constant bypass, the pump will always operate

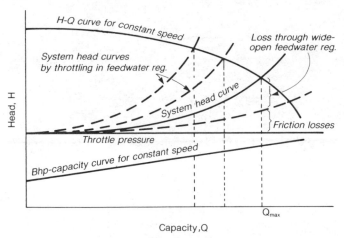

1.12

Capacity, Q

at greater flow than required by the boiler. An H-Q curve shows this (graph).

On the subject of controls: Since WR has two boilers, one of which operates 200 psig below the other, both boilers will have to retain their feedwater regulators. The simplest way might be to let the turbine governor vary pump speed to maintain a constant discharge pressure, allowing the feedwater regulators to throttle the rest. An excess-pressure governor to maintain a constant pressure drop across the feedwater regulators would give still greater power savings, but the controls would be too complex to be worthwhile.

I J KARASSIK, *Mountainside, NJ*

Can the old piping take new pressure?

WR's old pump can deliver 118,590 lb/hr of feedwater to the two boilers. The new pump delivers only 70,000 lb/hr to the two boilers, which is not enough capacity. Pumps should deliver 2 to 2½ times the steam rate in lb/hr of feedwater. The new b-f pump should be designed for the same discharge pressure as the old one, because the feedwater piping, valves, and regulators may not be able to take the new pressures indicated. It is a good idea to have a motor-driven and a steam-driven pump, so if he can use the exhaust steam, WR should buy a governor-regulated turbine to drive the pump.

J B WOODS, *St. Louis, Mo*

2 Pumps, Compressors, Valves, and Piping

2.1 Can money be saved by standardizing?

Our plant has several thousand steam traps of various types, sizes, and makes. Over the years, different departments have become loyal to a particular type or make—the boilerhouses follow their own selection, the steam-distribution system has another selection, the dryers yet another, and so on. Some equipment comes with the traps installed, and replacements tend to be duplicates of original equipment. Since all the lines handle steam, why can't we use one or two trap types and sizes to handle all the trapping?

Trap manufacturers give conflicting information when I ask for suggestions on selecting traps for different service. We are spending a lot on trap maintenance as it is, so apparently none of the proponents of the various trap types and trap-piping systems are anywhere near correct. What has been the experience of POWER readers with attempts to standardize traps? Did it save money? Are there any good general rules for steam-trap selection?—LM

First inventory, then compare trap designs

We encountered this problem in the powerplant and on about seven miles of steam-distribution and condensate systems. A multitude of traps had been installed over the years. Replacement and stocked repair parts were costly. Costs to maintain individual traps were difficult to compare. Also, in many cases, people associated with different parts of the system seemed to prefer different makes and types of traps. The problem is easier to solve than would appear at first, however.

The supervisor should get all the people together and ask them to take an inventory of all traps in service, to tabulate repair costs for the year, and to inventory the costs of the different types of parts for each trap type. Additionally, leakage and blowthrough should be estimated. In many process plants and distribution systems, these costs have already been tabulated, although no comparison has been made with different plant areas. In my experience, if this approach is taken, the inverted-bucket trap will win hands down.

Sizing of traps is next. Look at the bulk of services and find the capacities needed for them, then select one size for a fairly wide range of capacities, probably in the 30-70% range of trap maximum. In some cases, the orifice in the trap will have to be changed. Our experience is that on startup, a bypass may have to be opened for a few minutes if the system capacity is 80% or more of trap capacity. Range of capacity for the ¾-in. inverted-bucket trap is very large, probably larger than for any other size and type. It also functions at higher backpressures.

A large plant can get by with ¾-in., 1½-in. and 2-in. traps in most services. Additionally, traps of the same size can be installed at two different elevations, one above the other, for larger loads. For this, two ¾-in. traps would replace one 1-in. trap, so that parts would not have to be stocked for the 1-in. size. In cases where conditions change, as they do on a coil in cold weather, the two ¾-in. traps would give required capacity and also assure that additional condensate loads could be handled effectively.

S OSTAPOWICH, *Edmonton, Alberta, Canada*

A plant must have several types of traps

There is no reason why the vast majority of steam traps can't be standardized. Standardizing will cut the number of types and sizes of traps in inventory and reduce the areas of responsibility in dealing with factory reps of the various types of traps. LM should:

■ Get any one of the more reputable trap manufacturers to help in making a complete trap survey. Categorize for service, and note the level of repair— whether blowing through, leaking, not working at all, and so on.

■ After the survey, group all similar trap services. Those on similar service can generally be the same.

■ Decide on spending the necessary money for standardization. If LM has no formal trap-maintenance program now, it is a safe guess that a sizable fraction of the surveyed traps are wasting energy. Energy savings alone would justify the cost of standardizing.

In all likelihood, LM will need more than one type of trap, because process-steam demands differ from, say, freeze-protection steam-tracing demands. Although trap types cannot be narrowed down to just one or two, the trap brands can be narrowed down to just one.

R S DUBOVSKY, *Hopewell, Va*

Base decisions according to applications

Without knowing the plant steam system, it is difficult to be exact, but generally, you cannot expect two trap sizes and types to handle all trapping needs. A plant should not attempt to standardize according to type, size, or manufacturer, but the plant can and should standardize by application. This will save in initial purchase, repair, replacement, steam loss, and production efficiency.

First step in a program should be an independent survey of all the steam traps in the plant. You will want to identify pressure, temperature, location, type, manufacturer, backpressure, condensate load, dirt in the system, and

current status (whether failed open, closed, plugged, OK, etc). Divide traps
into applications:

■ Steam main drip.

■ Tracing, winterization, product protection.

■ Heating, unit heaters, radiators.

■ Process, reboilers, exchangers, coils, etc.

Then someone from the plant, or an outside consultant, who fully under-
stands the various types of traps available today, how they operate, and the
effect in your system, must select the type trap for the application category,
sizing it from system need. Examples of what can be put on main-line drip or
tracing services:

■ **1 psig to 25 psig to trap.** Thermostatic bellows, diaphragm (with subcooled discharge), bucket, and bimetallic all can operate with atmospheric discharge and low-pressure return discharge. All except the diaphragm can operate at discharge pressure over 20% of inlet pressure.

■ **25 psig to 100 psig to trap.** Choices here are thermostatic bellows, thermodynamic with low-capacity disc, diaphragm (with subcooled discharge),
bucket, and bimetallic. All these can operate with atmospheric and low-
pressure return discharge. All except the thermodynamic and diaphragm can
operate on return-line pressure over 20% of inlet.

For 400 psig and higher, consider the orifice, piston (impulse), and bimetallic. All three operate on all discharge conditions, except the piston on return
pressure over 20% of inlet.

J SCHMIDT and D KERR, *Houston, Tex*

It's important to keep traps in good repair

One of the most common misconceptions today is that one steam trap can perform well in all applications. You don't race a bus in the Indy 500 or plow snow
with a Camaro Z-28. Different types of traps are needed for best performance
in specific environments. There are other misconceptions, however. For example, the wish for a steam trap with the longest possible life, to give supposed
certainty that the energy-conservation effort will be maximized. Remember
that all mechanical devices will eventually fail, regardless of their quality.

Well-designed traps waste steam only after they have gone bad. In the long
run, all traps not under inspection and maintenance will waste considerable
energy after they have entered failure mode. The key to the best conservation
effort lies in identification and immediate repair. Incidentally, steam-trap life
relates more readily to capital expenditures for replacements than to long-
term energy conservation. The replacement cost is typically trivial when compared to the energy-loss estimate for a neglected steam system.

Trap life and quality are hard to assess, too. A particular model trap can
give 10 years of life in one application but fail in another after only three
months. This situation proves the need for multiple trap types, specifically designed for given tasks. Replacing all of Brand X traps by Noloss Corp's traps

and thereby cutting energy consumption by up to 40% is another misconception. If replacement of all traps decreased energy consumption considerably below the figures recorded when the plant was new, then indeed the plant has chosen a better trap. Usually, however, broad changeovers merely correct unattended trap failures and return the plant to a "like new" condition.

Because products of reputable manufacturers are almost always installed in this type of retrofit, energy savings will almost always occur. Nevertheless, the method is just an expensive maintenance procedure, hiding the need for an extensive maintenance program. Changeovers are to be recommended only when it is empirically determinable that poor traps originally went in. As a general rule, if serious maintenance programs are not introduced and followed over the long run, energy consumption will increase dramatically, regardless of the type, quality, or brand of trap.

What size traps are best for most applications? How many facilities have main drip traps and tracing traps larger than ½ in.? The undeniable truth is that almost all drip traps have loads below 200 lb/hr at 100 psig, and tracing traps have loads under 100 lb/hr; yet it is not uncommon to see oversized traps, capable of 2000 lb/hr, in these applications. Because traps don't perform well when oversized, these applications should be properly engineered and sized. A ½-in. trap can drain 300 ft of 8-in. 125-psig main with a 2-to-1 safety factor, at 0F and with 80%-efficient insulation. Main drips and tracers should therefore not be larger than ½ in. Specific-application benefits can be gotten if the following suggestions are followed:

■ **Drips.** Remove condensate as soon as possible, with little or no subcooling. A second choice is to subcool condensate if conditions tolerate or require subcooled-condensate traps.

■ **Tracers.** On critical service, remove condensate as quickly as possible, with little or no subcooling. For noncritical work, subcool the condensate to recover both latent and sensible heat of the steam.

■ **Process.** Rapid removal of condensate is desirable to prevent reduction in heat-transfer rates, corrosion damage, erosion damage, waterhammer, and stratification or temperature fluctuation. Install traps that discharge rapidly with little or no subcooling.

A common error in selecting process traps is choice of traps for pressure differentials based on line pressure. Although the trap rating should meet or exceed line pressure, this full line pressure is rarely found at the trap inlet. If traps are assumed to take full pressure drop but don't, then they are often undersized.

J R Risko, *Allentown, Penn*

Face the people problem, work toward a consensus

LM has a people situation to face first and, unless he handles that properly, whatever he does about traps won't help much. His men in the field have made choices for reasons, so he had better consult with his subordinates and work toward a consensus of opinion and an acceptance of a program. In that way, everyone will be part of the program and will make it succeed. Also, with input from the field, the engineering choices are more likely to be correct. LM cannot possibly be familiar with the local intricacies that the field people work with daily.

D L COCHRANE, *Sahuarita, Ariz*

Standardize end-to-end dimension

Replacing all the existing traps in LM's plant with one or two trap sizes and types would be analogous to replacing all the wrenches of plant mechanics with one or two sizes. There are too many variables associated with trap selection for the replacement to work. There *is* a system, however, that will cut downtime and improve trap maintenance. If the same dimension is held between unions or flanges for each size of trap, a replacement assembly can be piped in the shop. Then removal and replacement of a defective trap becomes a simple matter. No doubt, it will take several years to repipe all existing traps to standard configurations, but the end result is worth the effort.

D D COE, *State College, Penn*

2.2 Improve control on compressor drives

We operate four 1626-cfm, 100-psig, two-stage sliding-vane air compressors to provide for boiler and plant needs. Drive is by individual turbines, with 150-psig, 525F steam coming from our waste-heat boilers. Air load is for control and instrumentation on three fired boilers, three waste-heat boilers, reverberatory furnaces, pneumatic lifts at furnace doors, conveyor belts, and a hydraulic-pneumatic baling machine (about 40%).

Nearly all loads are irregular. The only air storage is at two furnaces and the baling machine—one 30-ft^3 receiver each, at the end of 700-800 ft of discharge pipe. Compressors load and unload 12 times an hour, with 310-hp peak draw per machine.

Present capacity control has a load/no-load pilot valve and piston arrangement, shutting the intake at 100 psig and reopening it when the system drops to 85 psig. When the compressor unloads and the intake valve closes fully, the turbine quickly speeds up, with the constant-speed governor attempting to control it. We think that the hunting action will harm the turbine. The turbines have one main and two auxiliary steam-admission nozzles. At present, all of a turbine's nozzles are always in service. Do POWER readers have suggestions for a better capacity control for us—keeping in mind that speed cannot be cut without reducing air pressure?—NCS

Control is not the only problem here

A governor to vary speed in response to air-pressure change will give NCS a good practical method of controlling capacity. Expertise in compressor turbine drives is necessary when a dependable, efficient control system is to be designed. NCS's system apparently has many more problems beside load control, however. He should be looking at:

■ **Air-storage capacity.** He needs adequate volume of receivers at the compressors.

■ **Air quality.** Instruments and controls need much cleaner air than the current installation sends to the plant.

■ **System safety.** "Floating" air receivers at line end, as described by NCS, are considered safety risks.

■ **Volumetric compressor output.** At 310-hp peak draw, the corresponding volumetric output at 100 psig is around 1400 cfm, well below the 1626-cfm rating NCS gives.

This compressor station needs considerable work, and the sooner the better.

S NOWACKI, *Parma, Ohio*

Go to sequential regulation on each machine

NCS's present control system is inefficient, and is likely to cause overspeed trip of his turbines. The alternative, sequential regulation and switching at load and unload pressure points on each machine, would cut cycling on the lead machine. The latter should unload just below the safety-valve setting, about 100 psig. This higher pressure setting should double the receiver storage time of one compressor. Cutoff points of the four machines should be 110, 100, 95, and 90 psig. Cut-in points should be 100, 95, 90, and 85 psig.

The turbine-nozzle opening should serve only to obtain required horsepower. Excessive governor throttling will raise the steam rate greatly. With the nozzle at maximum opening, valve movement and response time will be a minimum. The sliding-vane compressor is basically a positive-displacement machine, with discharge pressure determined by opposing line pressure. Speed can float to 60% of the manufacturer-recommended maximum, and some suction throttling is possible. One machine with variable-speed control sensitive to line pressure would give a 10% variation in capacity, and two machines so equipped would give 20% variation.

G A SCATKO, *Tampa, Fla*

Install an oil-relay governor for less droop

The governor control on the turbine may be the problem. If NCS has a mechanical governor, a change to an oil-relay type could help. The inherent inertia and speed droop in a mechanical governor are normally incompatible with a load/unload type of operation. An oil-relay governor, on the other hand, has a 1-3% speed droop, good for speed control, and response is rapid. Before

NCS makes any changes, I would advise him to:

■ Check the throttle valve for excessive friction, with the turbine at operating temperature.

■ Determine linkage conditions, and rebuild or renew if the linkage contains lost motion.

■ Dismantle the governor and check the terminal shaft and flyweights for ease of operation. Also try adjusting the droop.

If conditions persist after this, install an oil-relay governor, which is relatively inexpensive, simple, accurate, and sensitive. It is also more powerful than a mechanical governor of similar size.

D RADEBAUGH, *Saugus, Mass*

Hand-operate quality turbine valves

ˈFor a similar situation, in which our plant experiences reduction of air flow in stages from a high of 3100 scfm down to 400-600 scfm over two shifts, I plan to remove the steam-chest stops on the turbine auxiliary nozzles. Good-quality hand-operated valves will replace them. As the load rises or falls, we will open or close the valves in proportion to unit load, giving us some capacity control between minimum and peak loads. The pilot valve and piston can then control minor load shifts and maintain air pressure.

A CLARKE, *Jackson Heights, NY*

Get advance notice from acceleration

If the problem is considered basically to involve control of turbine overspeed during compressor unloading, then something similar to the anticipatory or acceleration governor controlling the overspeed during full-load rejection by a utility steam turbine can be fitted to the compressor steam-turbine drives here. The acceleration governor gives an overriding impulse to close the steam valves before the regular speed governor takes effect. The governor should

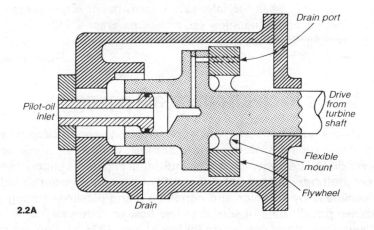

Drain port

Drive from turbine shaft

Pilot-oil inlet

Flexible mount

Flywheel

2.2A   Drain

close until the "no-load" limit of the compressor is reached. Impulse can come from either a mechanical or an electrical device.

The mechanical governor consists of a drive from the turbine shaft to a small, flexibly connected flywheel (see sketch, 2.2A). Pilot-oil pressure fed through the governor-shaft can be released if the turbine accelerates quickly enough so the flexibly mounted flywheel rotates relative to the governor shaft and aligns a drain port in the flywheel with the pilot-oil port. Release of the pressure then closes the steam valves.

V I KRISHNAMURTHY, *Shillington, Penn*

Put dump valves in turbine supply lines

NCS should check with the turbine manufacturer to see if the hunting action is acceptable. For turbine overspeed, 10% is within limits. If the hunting problem must be attended to, then either better turbine control or stage unloading of the compressors will help. Dump valves in the turbine inlet steam lines will dump steam when the turbine is unloaded (see sketch, 2.2B). These valves

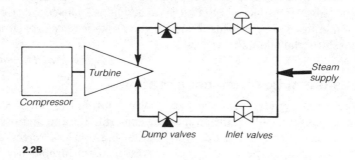

2.2B

should be set to open and close at 1-2% speed variation, and must be interconnected with the inlet valves, so the dump valves open only when the inlet valves close and close only when the inlet valves open. In our larger compressors, staging valves for unloading allow the turbine governor a wider time period and pressure range for steam reduction.

C R KLEIN, *Wauwatosa, Wis*

2.3 Ways to clean out steam piping

In our refinery and chemical-plant work, we often need to clean newly installed steam piping systems before startup. We do not agree among ourselves on the best way to do this. Individual engineers support several different methods, such as blowing out lines—with a target to indicate line cleanliness by its color and condition—and pickling, among other well-known possibilities. It seems to me, however, that all our methods have limitations. What can we do for large lines, 12 in. and up, for example, where high flow rates are needed to get cleaning-medium velocities that will carry foreign matter to an exit? Location and number of outlets and drains are other points that I would like to learn more about.

Can POWER readers give us guidance based on experience in cleaning steam lines, especially for low-to-medium pressure service? Is air superior to water or steam for cleanout? Will strainers help? If so, should they be temporary, or a permanent part of the piping? Should drain connections be permanent?—WJT

Swab and brush pigs precede high-pressure steam

Cleaning large steam lines can be a problem if large amounts of steam are not available. When we had to clean 7000 ft of a 24-in. and 18-in. 650-psig line, we turned down liquid-phase and vapor-phase chemical cleaning because of cost and effluent-disposal problems. Instead, we decided to pig the line with swab pigs (a plain polyurethane foam cylinder) and brush pigs (a denser polyurethane foam with strips of wire brush in the periphery).

First we rodded out the 90-odd drain boots, which removed some dirt and stones. We then pressurized the line with a little steam to blow out the drain boots. Before pigging, we dried the line with nitrogen. A pigging contractor then ran 36 sponge-rubber swab pigs and 15 wire-brush pigs through with compressed air. Each pig took about 10-12 minutes to travel the line. At the exit end, in addition to clouds of rust and mill scale, we found: wire-brush wheels, spray cans, pop cans, welding rods, and coupons. The cleaning itself was very successful. The final step was pressurizing the line with 650-psig steam, blowing each mud leg, and giving a final "target" blow with 100,000 lb/hr of steam.

D E TREAT, *Houston, Tex*

Cycle the steam to crack scale

All three methods mentioned by WJT are good within their limits. Air and water cleaning is normally restricted to small-size pipe, because large compressors and pumps would be necessary for large lines. For steam blowout, run the discharge piping to a safe location outside the building. Large volumes of steam are generally available for large steam lines because pipeline capacity and boiler capacity are matched. Several short periods of steam blowing alternating with cooldown periods are more effective than one long blow. Pipe expansion from steam heating will loosen and crack off mill scale on the interior. Temporary discharge pipes with brass targets to indicate cleanliness are needed where the lines connect to turbines and other equipment.

Condensate from a new system should not be returned to the boiler for several days. Test the hardness and iron content to establish the proper time. For water flushing, the supports and hangers must be able to take the weight of the pipe filled with water. Take precautions against water-hammer damage, too. A water-cleaning method that uses small volumes of high-pressure water could be an alternative.

High-pressure jet nozzles from a jet head remove mill scale, although the low volume of water will not flush out heavier objects. The jet head moves along the pipe by reaction force from the nozzles. This cleaning method has effectively removed calcium-based scale from process-liquid lines 6-8 in. in diameter. Contractors have this equipment available in truck-mounted units.

B M KINE, *Vancouver, BC, Canada*

Add compressed air to wire brush to remove debris

Here is a device that I put to good use several years ago on large steam lines (top sketch, 2.3A). A motor, connected to a wire-brush wheel by universal joint, is supported on adjustable spring-type legs for pipe 12 in. and up. Connect a compressor to the line first, and run a pull line and cable through to allow the device to be drawn through (bottom sketch, 2.3B). The wire-brush wheel will loosen foreign matter, and the air will carry it out. For sharp bends, a hinge-type wire-brush wheel is advisable.

J C LEIHGEBER, *Fayetteville, Ohio*

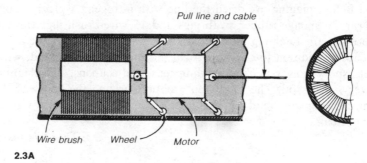

Pull line and cable

Wire brush   Wheel   Motor

2.3A

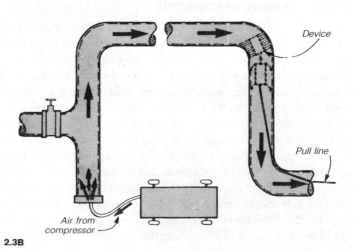

Device

Pull line

Air from compressor

2.3B

Follow air blasts with low-pressure steam

Compressed air is very effective. First, put quick-opening valves and targets at the discharge ends of the pipe system to be cleaned. Then fill the system with air to the maximum available or allowable pressure. Pop the quick-opening valve open and blow out the line. Read the target to see if more blowing is needed. In most cases, many blows are necessary, and the cleaning is an all-day task.

Low-pressure steam blowing cleans out grease, oil, and many other substances, and is usually done after air blowing. After steam blowing, remove the quick-opening valves and connect the equipment. Permanent strainers are a good idea wherever possible.

C R KLEIN, *Wauwatosa, Wis*

Take extra care in fabrication of pipes and fittings

Prevention is better than cure. In nuclear work, all large-diameter pipes and fittings are sand-blasted, and the ends tightly capped. Smaller lines are acid pickled to remove mill scale, dirt, and dust. After installation, the lines are cleaned by steam blowing, with a target to indicate cleanliness. Be sure to have temporary anchors and holddown arrangements. We have provided strainers only on the suction side of pumps not having a permanent in-line basket type strainer.

High-point vents and low-point drains are advisable. Low-point drains from the lower part of drip legs should be at least 2 in. in size. The drip-leg size depends on steam flow and quality of the steam. Each drip leg should have a bottom cleanout, too.

V S R KRISHNAN, *Houston, Tex*

2.4 Should dual-pressure air be used?

At our last review of plant-service costs, the compressed-air system received even more than the usual abuse. We in the plant engineering department realize that operation of our one 1800-cfm and three 2500-cfm compressors is costly, but we are unable to control end use of air enough to cut expenses by much. We are always checking for leaks, and have installed timer switches and spring-closed valves to cut waste. Now I am wondering if we should look again into a dual-pressure system, with most air at 40 psig and some at 110 psig. I know this has been repeatedly turned down in the past, but perhaps today's fuel costs have made it feasible.

Can POWER readers tell me what their experience has been with dual-pressure air systems? What are the best pressures? And what is the lowest-cost way to install such a system in an industrial plant?—ATM

Don't try in an old plant

Our plant had the same question come up last year. I took it on myself to look into the situation, and the results might be interesting to ATM and others. Incidentally, it cost us more money than you would think to make the study, and I caution anyone who assumes that such a study can be done in a day or so to rethink matters. I ended up by calculating air to pneumatic cylinders, based on displacement, pressure, and frequency of operation. Then I found air to instruments, valves, and controls, by metering air to the plant when no other activity was going on. At certain times, air use can be heavy for cleaning, and this flow was found as the added increment during cleaning periods, with a check made when the plant was shut down and only cleaning air was flowing.

We estimated that 4% of the 110-psig air went to cylinders and other high-pressure end uses, such as air motors. For controls, valves, and instrumentation, the percentage was about 23%. All the rest went for cleaning, cooling, and miscellaneous uses. For pressures, we found that most of our controls and actuators on valves operate at 30 psig maximum, and that for cleaning, pressure must be reduced even further. Header pressure necessary to squeeze the flow through the lines at their existing sizes must be around 50 psig, we decided. This allows a reasonable loss in headers at high flow rate, and still leaves enough pressure upstream of the instrument's pressure-reducing valves to give good control.

Finally, with all this information, we started to plan a better system. The existing piping could serve the low-pressure system—50 psig—but would still be marginal. An increase in load would require some added 50-psig piping in parts of the plant. A new compressor would have to be purchased also. For 110-psig air, we would have to run a new line, which could be smaller in size, but would still be expensive. The old compressor, running at greatly reduced load, would last longer but would be less efficient. The conclusion was that, while the dual-pressure concept was a good one for a new plant, it didn't make sense for us. We decided to continue with what we had, with added emphasis on keeping the system tight and stopping wasteful practices.

R HENSELMANN, *Omaha, Neb*

Two-stage compression is common

Compressing air to high pressure first, and then reducing the pressure to meet low-pressure requirements, represents a wasteful expenditure of energy in the extra compression, especially if needs for low-pressure air are large. Cooling-water requirements in the aftercoolers are also a factor. Don't forget, too, that the higher pressure means more leakage loss.

Two-stage compressors, with unloaders and sequential on/off controls for such applications, will give best results and eliminate pressure-reducing devices. Size the air receivers to correspond to the high-pressure and low-pressure needs, so unloading will be minimum. Savings in operating costs will compensate for increased cost of compressor and piping. The setup that I have described, with a low-pressure air receiver supplying the high-pressure compressor, is common in large power stations outside the US.

R K JAIN, *New Delhi, India*

60 psig is better for low pressure

In most plants, there is a tendency to add uses for compressed air, instead of trying to cut existing uses. Use of compressed air for people-cooling is one outstanding example of a wasteful practice, with all the energy in the compressed air disappearing in an inefficient cooling effect. Also, there are many "emergency" uses of compressed air on machines. Processing needs and cooling and clearing away of vapors are some of the reasons why a stream of expensive compressed air is used "just for a few hours." These hours stretch out, and the application becomes permanent, if no one stops it.

From my experience, there would be little benefit from a dual-pressure system, compared to the large benefit from converting as many end uses as possible to fans and blowers, portable if necessary. ATM mentions a pressure of 40 psig for the lower pressure of a dual-pressure system. This is below what I would suggest. Why not 60 psig, to give adequate margin for control systems and to allow him to connect some air cylinders to the low-pressure part of the system?

K F MARCONI, *Detroit, Mich*

Try decentralizing air supply

The idea of dual pressures for compressed air is worth close study for new plants. Actually, if ATM reflects on it, he may remember that many utilities do, in effect, have dual-pressure systems now, with high-pressure air for soot-blowing and lower-pressure air for station service. An occasional industrial plant also has a dual-pressure system; for example, a plantwide single-pressure system plus separate higher-pressure compressors serving individual machines.

New plants should look carefully at the requirements for any air pressures over those supplied by blowers. With the high prices for compressors and the high cost of electric power for compression, the stage has been set for big savings by careful energy managers. Our new process equipment is being specified to use less high-pressure air, with decentralized air supply in some cases. With low-pressure air, we expect much less trouble from moisture in the lines and from air leaks. In addition, by carefully figuring our air consumption and spotting small receivers at good locations, we see savings in piping, both in size and in number of lines.

E STANWICK, *Somerville, NJ*

Centrifugal blowers replace compressors

Our "dual-pressure" air system consists of 100-psig air piped through the plant, and several small, single-stage centrifugal blowers spotted where air output at 10 psig can easily reach areas to be cleaned. We have found that higher-pressure air is unnecessary for cleaning our type of lint; the small amount of really sticky stuff demands scraping and brushing anyway. We have stopped the use of high-pressure plant air for cleaning, and the power savings have been welcome. So far, we see no maintenance saving on compressors or drying equipment, but we still hope to show some benefit over the long term.

M R REYNOLDS, *Atlanta, Ga*

Put in a low-pressure line

Compressor manufacturers have been no help in advancing the worthy idea of dual-pressure systems. It is easy to see why. If all plant air, no matter what its final use pressure, must be compressed to top pressure first, the machinery will be larger and more expensive. Operating cost will be high, too. The only offsetting saving is in piping, but any plant with a compact layout can easily run the added line for lower-pressure air and still save money. ATM should look into the matter again. For an added cost-saving feature, he should recall that a shift of most of the compressed air to a new, lower-pressure line will allow the high-pressure line to operate at less pressure drop. The high-pressure compressor can then be cut back 10 psig or 15 psig for another saving in operation.

J S MANTON, *Albuquerque, NM*

Hose lines could be interim method

ATM is probably stuck with a single-pressure air system at present—like most of us. Here is one possible low-cost way to install a lower-pressure subsystem, however. The compressor for the lower pressure will be the most expensive item, but perhaps a used unit can be found. Then ATM could run some hose, designed for the pressure he mentions, around the plant, and see if the results indicate benefits. Moisture removal from the lower-pressure air is not so important as for high-pressure air. The equipment requiring low-pressure air is often not sensitive to water droplets—although instrumentation is a clear exception here.

C C BURCHARD, *Des Moines, Iowa*

2.5 Protecting small piping from damage

The large piping in a plant may cost more than the small lines often referred to as "spaghetti," but our maintenance section has more trouble with small lines than with large. In the reinforced-concrete building of our plant, we have tended to follow the original routing that the plant designers and contractors selected—against columns and close under beams. We have had heavy damage to small-size piping from such causes as fork-lift trucks, carelessly handled scaffolding, and swinging hoist loads. We have tried guard rails near ground level. They help, but we cannot guard enough of the line length to prevent continual repairs.

We now try to run our instrument tubing in troughs and racks, but some lines must be in the open for part of the run. We should improve our routing or protection for small water, air, chemicals, and drain lines, too. Steel enclosures tend to make installation and repair difficult, but we may go to them yet. Channel shapes might be an answer, but maintenance can be a problem with such a layout. Valves, too, are difficult to install and service when the line is enclosed in steel. We use very few welded fittings on small lines, for this reason. What has been the experience of Power readers with protection of small lines? Should we run them in a protected location, or is it better to install removable guards? — RHU

Owners, be on the scene early, improve liaison

RHU's problem is a common one, and little other than the measures he mentions can be done to protect the piping once it has been installed. The problem is getting worse with the wider use of less-rugged plastic piping and tubing. Normally, when plants are designed, the emphasis is put on the large piping, which must be defined early to get the spool sheets to the job to meet schedules. The small piping is left to the less experienced designers with little field or operating experience or often to field forces, whose piping crews run it to suit. Often the crews are the only ones it suits.

Owners can improve this by developing more liaison between operating personnel and designers, in either their own or contractors' design offices. Early appearance of owners' field personnel during construction can correct some of the more glaring examples before they become too expensive to change.

G S Coblentz, *San Mateo, Calif*

A removable guard can help

Best way to protect small-size pipes and tubing from being damaged by fork and hi-lo trucks is to move the piping or tubing to the opposite sides of the columns from the aisles. This should prevent the trucks from getting to them. In some areas where piping is being continually damaged, removable steel guards should be installed. RHU can't install guards over all piping on columns because the cost of making and installing the guards would not be offset by repair savings.

To stop the damage, an educational program should be begun for truck operators and handlers. A course in plant safety and material handling is advisable. When the course is finished, the operators should get a permit or license to operate their equipment. Later, if they are involved in any accidental damage to building or equipment, the permit or license should be revoked for a time. Perhaps during that time they could be given a lower-grade job .

G B Hill, *Sunbury, Penn*

Steel channels solve the problem

We have gone to large steel channels for our instrument lines and find this to be satisfactory. We run the channel on a protected side of a column and then put the instrument lines in it. They are flexible enough to be drawn out far enough for work at joints. We stagger these joints, of course. For horizontal runs, we insist that piping be high enough to clear trucks, at least when forks or cranes are down for transport.

J L WARLETT, *Baltimore, Md*

Color-code lines at fittings

First priority should be to modify the operation or educate the workers to exercise more care. If this is impractical, take a close look at the lines and the structure through which they run, from origin to final destination—maybe something has been overlooked. For vulnerable areas, a type of tubing-channel protection is best. Avoid suspended ceiling areas whether protected or not. Try to make fittings accessible by notching the sides of the tubing channel. Try to locate fittings on nearby lines close to each other to make troubleshooting easier.

Vibration reduction will help keep fittings from working loose, too. On the subject of fittings, follow vendors' recommendations to the letter where tightening is involved.

For running tubing across concrete, there are many effective concrete anchors, and a tubing vendor can help here. Don't forget to color-code the lines at the fittings, either. Many of us have traced a line in a channel for 50 ft or so, only to have to start over when we found it was the wrong line.

If a massive repiping program is decided on, the most sheltered route through which many lines can be run is the best. The additional tubing material is expensive, but the consolidation of many runs, with separate supports, channels, and scaffolding setups for workers, will pay off. An experiment with some bothersome lines could show the right direction.

For the last question, I would avoid removable guards and would strive for fitting accessibility where possible. A removable guard is a big initial expense, and it will disappear because it will not be reinstalled after later piping repairs.

A D CONRAD, *Elizabethtown, Penn*

Run along walls of pipe trench

Small-size piping should be run overhead, so that people cannot walk or sit on them. Where piping comes down from overhead, it should run outside of columns but attached to the column. In a building where that would be unsightly, the piping can be run in pipe chases or inside dummy columns. Small piping underfoot should be placed in a trench or on the walls of a pipe trench, covered by floor plates or grating.

Pipe that comes up out of the floor between columns must be protected by an enclosure, such as angle, to which it can be attached to keep it from bending or being run over. The same kind of protection serves for piping that drops down from the overhead between columns.

J RAMPOLLA, *New York, NY*

Trusses and trays combine

Accidents to piping runs such as described by RHU can be expected. I have even seen a pipefitter drop a wrench through a newly glazed window while the glazier was fitting the next pane. Cranes, lift trucks, and even hand-carried ladders will cause the damage described. My people had kindred problems with piping, tubing, and electric power and signal cables. To solve the problem at least partway, without armor-plating or relocating the runs, and yet keeping them accessible, we resort to U-shaped trays high on walls and carried by brackets. The trays cross roads and passages as expanded-metal formed trays welded to rebar-type welded roof trusses that span 20 ft or more.

Our runs included hydraulic and air tubing of metal and plastic. The cables include plain wiring and shielded cables, all easily accessible and high enough to be above trucks and lifts in the area. Only in the most heavily traveled areas did we go to underground tunnels. There we used trays again, but with underneath clearance for drainage and heavy steel covers easily removable for servicing.

Trusses and trays (cableways) are available from many building-supply houses. The expanded tray welded to the top of the truss gives extra rigidity to the span. For further protection, we painted the tray sides exposed to traffic in brilliant orange or yellow for visibility and warning. For electric cableways, remember to lay a bare groundwire the length of the tray system, and weld or braze it to the trays at intervals. This will completely ground the system.

H M NEUHAUS, *Santa Monica, Calif*

2.6 Valves needed on condensate lines?

Our plant requires steam for heating the building, process water, and drying air. Nearly all of our operations shut down over weekends, during which time only the heating load, about 40% of the total in winter, is on. We generate header steam at 125 psig and return about 75% of our load as condensate. We trap all our equipment, of course, and we have been installing drain and return piping pretty much as recommended, with isolating gate valves, check valves, bypass lines, and strainers at the traps.

Lately, I have been looking into reasons and results in connection with the drain and return piping, and now I am seriously considering elimination of all shutoff and check valves in drain and return lines. I find that most of these valves leak badly within a year after installation. They do not shut off tightly enough to let us work on piping or equipment until weekends. Our piping installation and repair costs are increased considerably by adding the auxiliary valves near traps, and I think that, if we eliminate these valves, our overall expenses will be reduced. This proposal is being fought by several men in our organization and by some vendors, so I would like advice. On the basis of their own experience, what do Power readers think of my plan? Are there any other ways to improve the operation of our drain and return piping system?—BL

Power Piping Code says 'Yes' to valve use

In considering BL's proposal to eliminate shutoff and check valves in steam-trap return lines, first priority must be given to the Power Piping Code, ANSI B31.1, paragraph 102.25, which states: "Where several traps discharge into a single header, which is or may be under pressure, a stop valve and a check valve shall be provided in the discharge line from each trap." If the system has been designed, or can be modified, so that it cannot be pressurized, valves in condensate lines should be eliminated. Such a system would require a condensate tank at atmospheric pressure.

Here are two questions that should be asked in deciding on the installation of a bypass valve around a trap:

■ Do you need a faster warmup of the system than the trap alone would provide?

■ Can the steam-using device be taken out of service for trap maintenance? (Traps at building heaters, for example, do not usually require bypass valves.)

A block valve upstream of the trap is, of course, a necessity. It should be kept in good working condition, too.

C G Day, *Avoca, Mich*

Check into water treatment first

If valves are becoming too leaky to isolate traps within a year, BL is using improper valve material for the job. More exotic valve materials cost more at first, of course, but reduced replacement labor will repay the initial expense. Surely, weekend repairs at overtime rates are not cheap maintenance. Use of the more dependable valves will allow repairs to be made during operation. A decision on whether check valves are necessary between trap discharge and condensate return would be based on the effects of back pressure on the traps. I question the need for continued use of strainers after the initial plant or line startup. BL's experience with foreign material in his traps can guide him in making that decision.

In my opinion, based on experience since 1937, leaving the valves out both ahead of and after the traps is inadvisable. The vendors naturally have their own axe to grind on that score but, even so, a leaky valve is better than nothing when a trap body or gasket fails and live steam escapes all over the area. Just one such incident requiring a plant shutdown will pay for many valves and manhours. One last suggestion: Investigate the feedwater quality to find out why the valves are leaking in such a short time on condensate. The pH level should get special attention here.

B B Brown, *Hagerstown, Md*

Strainers and check valves can go

Because of my similar experience with valves at steam traps, I share BL's opinion to a large extent. Before he eliminates these valves, however, I think he should evaluate equipment for importance in process and for loss of money caused by equipment downtime while the trap is being replaced. Superquality valves to bypass these traps would pay off.

Check valves in condensate drains to return lines help prevent backflow of condensate to standby equipment if the condensate line has become overloaded. Other than that, they are more trouble than they are worth, and I would get rid of them. In regard to strainers, if the water treatment is good and there is no corrosion of steam and condensate return lines, there will be no trouble if the traps have no strainers. If the traps are hooked up as shown in the sketch. BL should be able to get rid of most of the strainers and bypass valves around the traps.

S Ikramullah, *Des Plaines, Ill*

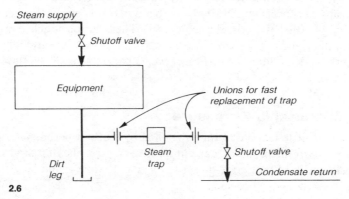

2.6

Main steam traps need shutoff valves

The valves on condensate and return lines are in a difficult service, certainly, and BL is not the only person who has had trouble with them. Over the years, I have seen a trend to fewer (if not better) valves on these lines, and I heartily concur with this. I am not sure why the valves on this service don't last as long as we would like, but I suspect that the main reason is the two-phase or alternating-phase fluid flow through them. They get steam with water drops and slugs, water with steam, flashing water, various temperatures of water in rapid sequence, and some dirt and scale. Is it any wonder the valves give up in a few months?

Flow of condensate through piping and heaters tends to loosen scale and dirt, which is then carried down across the seating surfaces and may help wash them away. We tend to rely on strainers to prevent this scale and dirt from getting to the trap, and I think this would be a wise precaution for BL, even if he discards much of the valving. We put small gate valves on the strainers, for blowdown. These valves don't give us any trouble, because they are closed all the time. Dirt legs in lines from equipment to traps are another spot where a blowdown valve is useful, because these legs seem to fill up quickly with sludge and dirt.

Bypass valves are one type that we have eliminated, and I note that several trap manufacturers don't show these in their layout schematics either. Other plants must have had trouble like we did. The shutoff valve ahead of the trap we believe to be unnecessary in most cases. Nearly all heaters and other equipment have a steam-shutoff valve ahead of the unit, and this valve is in a less-severe service than the condensate-line valves and lasts longer. We rely on it to isolate traps.

Valves downstream of the trap are another headache. We have tried several types here, including ball-valves, globe valves with plug-type discs, and gate valves, but we have had little long-term success with any. Invariably, when the time came that we needed a tight shutoff, we didn't get it. The check valves often installed downstream from the trap are another disappointment, as far as tight closing is concerned. Draining of steam mains is one case where shutoff valves are justified, BL should remember. There is too much danger from water hammer and equipment failure not to have a shutoff valve ahead of the traps on these drain lines. In this case, even a part closing of the flow could be very necessary.

J R SIMMONS, *Portland, Ore*

No problems at low pressure

We have had little trouble of the type cited by BL, but perhaps the reason is the low pressure of much of our equipment. We run heaters and dryers at 40-50 psig and find that good-quality bronze gate valves hold up well on condensate lines. We have been adding amines to our boiler feedwater for over 10 years, and this has helped avoid corrosion.

H PIETROWSKI, *Springfield, Mass*

2.7 Heater pressure rises when valve opens

A 30,000-lb/hr heater is giving us a problem. The heater operates with a 1½-psig pressure, as shown in the drawing. The multiple-port valve relieves pressure at 2 psig, but when it does so, the water level rises in the heater until it reaches the overflow. Then pressure in the heater rises to 10 psig. This pressure holds, with nothing leaving the exhaust head, until the water drains from above the multiple-port valve. This takes 15 to 20 minutes. Then everything is normal again. Water level in the heater ordinarily varies from half to three-quarters of the gage glass. The PRV closes tightly at 2 psig and does not leak. I once shut off the makeup-water valve, but there was no change. Pressure over 5 psig on the heater will prevent water from flowing from the condensate tank.

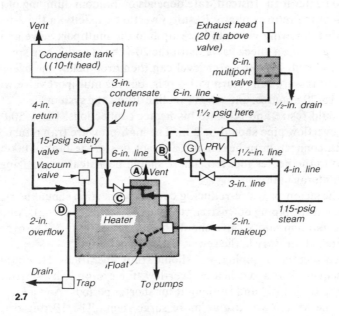

2.7

Recently, with heater level 2 in. below gage-glass top, 10 psig in the heater, and no water from the overflow, I opened the safety valve and found hot water. Perhaps it was forced up into the 6-in. line. Can POWER readers give me clues on what is happening here? Why does water collect above the multiple-port valve, and what should I do?—EH

Connect relief line to heater, add level-control valve

If the 6-in. steam-supply line and the 4-in. return enter the heater below normal operating level, then perhaps when the multiport valve opens, the heater pressure forces water into the 6-in. steam line to a level that balances heater pressure. The steam trapped in the heater could then maintain the 2-psig pressure while the 4-in. return continued to bring water to the heater.

The rise in level will increase the pressure and force water into the steam-supply line and to the relief valve. Water from the condensate line would not increase heater pressure, but would condense trapped steam. An overloaded return and/or a cut in HW demand could cause the pressure rise. EH should remove the 6-in. relief line from the steam-supply line and connect it to the top of the deaerating section of the heater, near the vent (A in drawing). He should also put a level-control valve on the 4-in. return.

B H RIUTTA, *Green Bay, Wis*

Trap blowby causes the 10-psig pressure

In EH's system, condensate returns via the 4-in. and 3-in. pipes at a rate unrelated to heater load. Instead, rate depends on random dumping of traps and operation of the returns pump. Result: Overflow capacity of the 2-in. line can be exceeded, allowing water to back up into the multiport valve and beyond. The valve acts like a check valve, with the 20-ft head and 2-psig spring setting keeping it closed. Heater water level can then drop below the overflow pipe, and heater pressure can return to 1½ psig, but the multiport valve will still be filled for 15-20 minutes, until the ½-in. line drains the system.

EH should resize and relocate his heater connections. For a 30,000-lb/hr unit, the overflow pipe should be 3 in. The high-pressure trap return would be 2-in., with connection in heater top and not in storage tank. Makeup steam should go to the heater top, too. There should be a steam-equalizing connection to the storage tank.

Also, the steam-pressure-reducing control connection should be right at the heater, not at the piping downstream of the PRV (B in drawing). And the pressure gage and pop safety valve should be on the heater top. The overflow line should pitch all the way to the sewer, without pockets.

After correcting the piping, EH should study operations. He might arrange returns sequence to avoid heater overflow, for example, by starting returns pump more frequently and running it for shorter periods. And a lower level of water in the heater would give more surge room. The 10 psig occasionally found comes from trap blowby; thermostatic and impulse traps are possible causes. Don't waste this steam, EH; set the multiport valve at 5 psig and leave the PRV at 1½ psig. Then, if pressure ever exceeds 5 psig, go to work on your traps. If you can't hold pressure under 5 psig, raise the multiport setting to 10-15 psig and put in a small condensate pump to help the gravity feed. And finally—don't overlook leakage of the 2-in. CW makeup. If 35-psig water is handy, a ¾-in. valve will supply enough water and cut the shutoff leakage.

M A MANGINO, *Boston, Mass*

Steadier hot-water flow is best at all times

Two assumptions are necessary to get a solution to EH's problem. First, the heater condenses steam, which immediately suggests a dip pipe on the inlet steam. A dip pipe can explain why water is at the safety valve with no water in the overflow, and the heater level 2 in. below the gage-glass top. Second, I assume that the problem starts when the user stops or sharply cuts his HW flow. Then the return-line water could drain back into the heater. If the user is higher than the exhaust head, problems will really start. With enough water in the return line, the heater's water level would continue rising and build up pressure to vent through the multiport valve. Vapor would be displaced from the heater until the returning water reached the level of the inlet dip pipe. Then the water would back up into the dip pipe, the 6-in. header, and 20 ft or so to the exhaust head. This would approximate 10 psig on the heater.

When the user calls for hot water again, everything will return to normal when the return pipe fills with water. If I am right, then the best solution is to maintain a steady HW flow at all times. An automatic bypass valve around the user is one idea. Also EH should check the condensate tank to see if it will hold the amount of water in the 4-in. return header. If it will, remove the internals from the check valve in the 3-in. condensate return (C in drawing). Also, relocate the 6-in. multiport valve closer to the exhaust head and higher than the condensate tank. Then, during an upset, the water will fill the condensate tank instead of the 6-in. line, multiport valve, and exhaust head.

E STECK, *Mt Vernon, Ind*

Intermittent returns need space

A 4-in. overflow valve located at three-quarters of tank height should be put on the heater. Then the makeup sensing point should be set at one-third of heater-tank height, to give enough volume for intermittent returns. The vented condensate tank should not drain to the heater; instead, a condensate pump sized to run at least 50% of the time should be installed. The condensate should never be held up. Relocating the multiport valve as close to the heater as possible, with free discharge, would also help, along with a 1½-in. drain from it.

I A KAYE, *Southington, Conn*

Send returns to condensate tank to improve control

I think uncontrolled condensate flow into the heater is the cause of EH's trouble. From the diagram, the condensate tank basically operates "dry," and all surge capability of the system rests in the heater tank. When half to three-quarters full, the heater tank has little surge capacity, so overflow conditions could exist there if steam flow fluctuates widely. The 2-in. overflow line and trap are inadequate. Try this:

■ Relocate the 4-in. return line from the heater to the condensate tank.

■ Install a 4-in. overflow line from condensate tank to drain or to heater tank.

■ Put a second float valve in the heater below the present valve, and reconnect the 2-in. makeup to it for emergency.

■ Reroute the 3-in. condensate line through the present float valve in the heater to control condensate flow to heater.

P D WEBB, *Richmond, Va*

Try an overflow loop connected to heater-tank opening

Build-up of heater-tank pressure can result from the 3-in. return if condensate-tank vent is blocked or inadequately open, but excess pressure from the 4-in. return is the probable cause. Correct this situation first. Then remove the 15-psig safety valve and the 2-in. trapped drain. Next, a 6-in. over-flow loop should be connected to a 6-in. heater-tank opening (D in drawing). Allow 27-in. leg height for each 1 psig of maximum pressure desired before loop blowthrough. Put an antisiphon nipple at loop top bend, and a refill water connection for use if the loop blows through.

M B GOLBER, *Chicago, Ill*

Check the multiport valve drain, add storage tank

With only a 2-in. overflow and 3-in. and 4-in. inputs, the heater tank could be overloaded. The trap in the overflow is perhaps the more usual overflow of a shutoff valve operated through a mechanical linkage from a float. The $\frac{1}{2}$-in. drain may be clogged if it requires 15 to 20 minutes to drain the water from above the valve. The basic trouble, however, seems to me to originate ahead of the 4-in. return. A high flow of condensate entering faster than it can flow out of the heater would fill the tank. The multiport valve will open when 2 psig is reached, and then will remain open as pressure rises because of static head of water. I suggest a storage tank, with overflow, for the 4-in. return line.

A M PALMER, *Brooklyn, NY*

2.8 Checking pump performance easily

Keeping track of the several hundred centrifugal pumps in our plant is a difficult job. Recording the basic characteristics and location of each requires a card file and several manhours a week of follow-up time. Pump performance is an additional area that we would like to check, but we know of no easy way to do it. We rely on what the manufacturer tells us about performance curve, NPSH and necessary drive power, but we cannot always be sure of what we are getting—from either a new or an in-service pump.

Several months ago, we installed a pump with a specified impeller diameter to give us a definite head and capacity for an existing service. We spent several hours trying to readjust controls and valving before we decided to open the pump to look for the cause of the trouble. We found an undersize impeller that made the pump's performance marginal. Perhaps we could have found this immediately on startup, or even before, but we don't want to have to dismantle every new pump or install a test setup.

What experience have POWER readers had with checkout of pumps on-line? What elements of existing instrumentation can we rely on for clues to performance and condition? And how can we judge pump condition without opening the pump?—NRE

Shutoff head reveals internal wear

With a characteristic curve and a knowledge of impeller diameter, both obtained from the manufacturer, NRE can make an approximate-performance test as follows: Take pressure readings at suction and discharge to find total net head being developed. Take ammeter reading on motor to find power. With these data and the characteristic curve, follow the impeller-diameter curve to the intersection with the horizontal line, indicating total net head developed by the pump. The horsepower requirement calculated from the motor load will further verify this point.

Internal wear increases clearances and decreases discharge head for a given capacity. If a pump has a valve in the discharge piping, a shutoff-head test will show if excess wear has occurred. Compare suction and discharge pressures to get the total net head. This shutoff test is also a quick way to check for proper impeller diameter.

J SLOWIKOWSKI, *Erie, Penn*

Field test demands precautions; changes stressed

Performance of a pump can be found from a test covering capacity, total head, pumping temperature, speed and pump bhp. A field test imposes certain difficulties, however, because only seldom can a pump installation meet standards for precision. Probably the least accurate measurement in a field test is that for power consumption. How accurate need a field test be? If it is to demonstrate that a pump has met its guarantees, it is doubtful that the accuracy will be good enough. A witness test at the factory would be better. If the field test is made to establish the degree of deterioration in performance because of internal wear, the problem is even harder to solve.

Internal wear's effect is to short-circuit part of the pump capacity to a lower-pressure area in the pump, reducing net capacity. Leakage will vary approximately with the square root of pressure differential, and so will not be constant for all heads. Assume that leakage is constant, however, and the result will be a constant amount of capacity deducted from the values at all pump heads. The general recommendation is for an overhaul when pump effective capacity is reduced by 4-5%. A field test, however, is at best accurate to only within 3-4%. Obviously, a field test cannot be an effective guide for overhaul need unless you take specific precautions.

Errors in measurements must remain consistent—changing little in magnitude and not at all in direction. It is *changes* in capacity, head, and power that interest us, rather than the exact values. There are several possibilities for a plant operator to determine degree of wear. Assumption is that a test has been run shortly after installation, and that the same instruments and method are used for the retest. If the pump is constant-speed, measure flow for a given discharge pressure, with unchanged suction pressure, at or near rated conditions. If the pump is variable-speed, measure rpm needed for a specific capacity and total head. Convert the data to full-speed operation, and compare with original test. Measure motor amps needed to meet a certain capacity. Compare with readings after initial operation. If the pump has an axial thrust-balancing device, measure the balancing-device leak-off. It will usually increase at the same rate as leakage past other internal joints. When leak-off has doubled, the internal leakage has probably doubled, and clearances should be restored. If the pump has sleeve bearings, remove the bearing bushings, packing, and seals, and lift the rotor with a pinch bar. Comparison of total lift at initial operation and after several years will give a good indication of clearance increase. Basic approach should nearly always be: "Don't open! Let it run!"

I J KARASSIK, *Mountainside, NJ*

Head/capacity curve gives clues to performance

To find the most common mistakes, all that NRE needs are the expected performance curve for the pump, a suction pressure gage and a discharge pressure gage. (If the driver is not an electric motor, a speed measurement is necessary, too.) After normal checks, such as for priming and rotation direction, close the discharge valve and read suction and discharge pressures. Don't operate the pump at this condition for more than a few seconds, however, to prevent overheating.

Then open the discharge valve in several steps, and read pressure at each step. Even without flow measurement, this will give a rough idea of shape of the head/capacity curve. Compare it with the expected head/capacity curve. If shutoff head is lower than expected, and the head decreases with increasing valve opening, chances are that either the impeller diameter is too small or the wearing-ring clearance is too big. In an open-impeller pump, it may be the impeller-to-casing distance that is excessive.

Low shutoff head, with head increasing slightly as the valve is initially opened, could be caused by an impeller installed backwards. The three possible curves look like those shown (2.8A). If shutoff head is at or near expected values, but pump performance is substandard at normal flow conditions, then no easy check is possible. Partial blockage of waterways in pump or piping, insufficient NPSH, or system resistance higher than calculated could be causes.

J BIHELLER, *New York, NY*

Pressure-gage method is useful, least expensive

I find the least expensive and most useful method to be one based on pressure gages. For initial testing, gages P_s and P_d are installed, along with a temporary flowmeter and differential-pressure readout—unless a permanent metering device is installed. Instruments to measure fluid temperature and pump driver power are also part of the setup. Gages and meter look like those shown in sketch (2.8B).

In tests, P_{d1}-P_s establishes pump head, which must be corrected for fluid temperature and density effect. This, with the flow measurements, allows a check on the manufacturer's performance curve and NPSH requirement. Input power readings permit us to verify efficiency. After this, if no permanent flowmeter is provided, the temporary orifice comes out, and gage P_{d2} goes in downstream of the throttling valve, elbow, or other restriction. Then, the curve of $(P_{d1} - P_s)$ vs $(P_{d1} - P_{d2})$ is found. The pressure drop $P_{d1} - P_{d2}$ across the elbow or valve is subsequently used to measure flow. Gage P_{d2} can be removed, to be reinstalled only when a check is necessary.

N R STOLZENBERG, *Suffield, Conn*

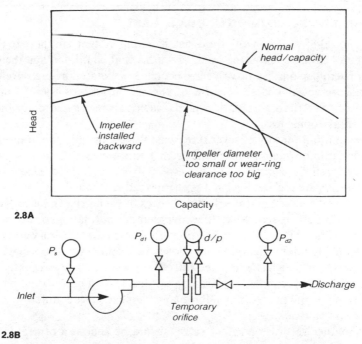

2.8A

2.8B

2.9 Safety-valve maintenance intervals?

We operate a large chemical plant, which requires high-pressure steam. We have many safety and relief valves on various services. Up to now, we have followed a program of removing and maintaining these valves yearly, or as close to yearly as possible. Our records are not complete on findings during inspection but, to the best of our maintenance men's recollection, the findings on internal condition have not been the same from year to year on valves in the same service. One year the men would find a relief valve dirty; the next year it would be entirely clean.

Safety and relief valves seem to leak at random, too, and we have been unable to predict how long a valve will last on a given service before it begins to leak. We have heard of attempts to schedule maintenance and inspection of safety and relief valves in a way that makes use of inspection findings to determine the time to the next inspection.

We wonder if this scheduling could apply to our plant. Most of us would be skeptical of any inspection and maintenance program different from the yearly-interval one we now follow. Have Power readers had experience with attempts to stretch out the interval between overhauls? If so, how can we decide how much longer a valve can go without removal and dismantling?—LDV

Systematic, scheduled check is best

Safety and relief valves in any fluid/gas-handling system are there to protect the system against unexpected loads or shocks that could damage the equipment or harm persons. Some valve inspections are dictated by pressure-piping and pressure-vessel codes, because of the seriousness of these protection demands. Other than automatic signals, such valves are the main bulwark against catastrophic loss.

Valves can fail for any of several reasons: Wrong installation, internal corrosion, buildup of deposits. Remember that rupture diaphragms, which are backups for relief or safety valves, are especially susceptible to these effects—more so if they are old and have not been replaced or cleaned.

The answer is a systematic, scheduled check of each valve in place—merely lifting the disc off its seat while the valve is under normal operating pressure and noting whether the valve seats properly. This is satisfactory for interim checks, and a regular schedule should be set up for this. Any valve that doesn't seat properly, or that has a damaged part—such as one I saw recently where the hand lifting lever pulled off because the disc was stuck—should be pulled and replaced. If the plant is down for maintenance, then the same valve can be repaired and put back.

Maintenance records on safety valves should be kept at a central location. Test-date schedule should be rigid, not only for safety's sake, but also to satisfy insurance inspectors and government authorities in case of surprise inspection or casualty investigation. Company engineers should check each valve's setting, even if the valve was bought under specification from a reliable supplier. I recall a shipbuilder who set up a valve test bench in his shipyard because suppliers were not conforming to specs even though the valves were certified. Tagging safety valves with valve number in the system, date of last overhaul, pressure setting and date of last inline test is a good idea, too. These data should also be entered on the central records, to protect against loss of valve tags.

H M Neuhaus, *Santa Monica, Calif*

Recommends staying with yearly maintenance interval

When LDV speaks of safety and relief valves, I assume he means valves in service on process equipment, not valves on boilers. It is true that stretchout of maintenance intervals on these safety valves is theoretically possible, but I question the wisdom of spending time trying to lengthen out a single maintenance interval such as this, when many other aspects of maintenance could offer a more profitable target.

To start with, a yearly basis is a very good one, because it agrees well with plant operations. Turnarounds are often on this interval, and coincide with slack periods in production. This means that a routine can be built up that is easily remembered. To change this to some indeterminate interval, or to an interval of several years, is asking for trouble. Memories are too fallible, and there is too much chance for error, even with a supposedly foolproof reminder system.

Changes in processes are another reason why I would not recommend deviating from the annual inspection and teardown. All the careful records in the world will not be useful if a process variable or a corrosion condition changes during the year. These changes occur constantly in plants and affect all the equipment, including the safety valves. Some of the stretchout systems that I have seen rely on data collected over several years of valve use before a change from annual inspection and teardown is advised. This is too long a period to be worth the wait, I say.

R CUMMINGS, *Philadelphia, Penn*

Keep a record of each safety and relief valve

Without a good, complete record-keeping system, LDV will not be able to do any reliable predictive maintenance on his relief and safety valves. Mechanics' recollections are good for general information, but not for specifics. Some states require annual inspection and testing, particularly on power boilers and other coded applications. Any attempt to stretch out inspection schedules must be viewed with caution.

Other services may have their inspection frequencies extended if it practical to do so. A time-tested method in some plants is to keep a record on each safety

```
SAFETY VALVE INSPECTION RECORD

Valve location_____   Identification no. _____

Manufacturer _____    Mat'l of construction_____

Set pressure _____    Discharge rate_____

Type of valve _____    Material relieving_____

Valve condition at inspection:                    Date_____
Clean_____ Leaking_____ Relieving pressure
Dirty_____ Not leaking _____ before adjustment_____

Date of previous inspection_____

Remarks:
Adjustment required _____
Repairs required _____
```

2.9A

and relief valve, like the one shown in the sample (2.9A). The data can be put on a computer if desired. LDV should realize that it will be some time before his records start to tell him enough to reschedule or extend inspection periods. After some time has passed, however, the frequency can be extended in some cases, and then labor savings will build up.

B B Brown, *Hagerstown, Md*

Put safety before economy when testing

Perhaps these safety-valve practices will help LDV: First of all, the maximum operating pressure must always be below the pressure at which the valve re-seats. The margin between operating pressure and the popping pressure of the lowest-set safety valve should be as large as possible, but not less than 5% of the set pressure for pressure below 1500 psig, and 7% for pressure over 1500 psig.

When LDV is setting up his testing criteria, he should consider both safety and economics, but safety considerations should *always* take precedence over economic ones. For a suggested testing schedule, try this:

■ For steam pressure of 16 psig to 900 psig, test manually once a month, and pressure-test once a year.

■ For steam pressures over 900 psig, test manually every six months and pressure-test once a year. Complete disassembly for inspection, cleaning, and testing is a good idea.

Operating experience is the basis for determining what the testing interval should be to keep safety valves in satisfactory condition. Some of the more common troubles that affect the life, operation and efficiency of safety valves are:

■ **Leakage.** A safety valve that hisses or leaks at operating pressure could be the result of damaged seats, tight lifting gear, distortion or wrong operating pressure.

■ **Incorrect popping pressure.** It can be too high or too low.

■ **Blowdown.** The normal figure is considered to be 3-4%.

A J Spencer, Sr, *Columbia, Penn*

Watch the feedwater treatment; valves are affected

A safety valve should be checked daily, either by overpressure or by hand, to make sure that it is in working order. Dirt under the seat is a big cause of safety-valve leakage, so chemical treatment of boiler feedwater is one place that should be checked to make sure that incorrect treatment is not responsible. If a safety valve should leak after popping, lift it by hand to drive out any dirt.

Drains from safety valves should be looked at closely, too, to make sure of the pitch. This will keep water and dirt from remaining on top of the disc. Another preventive-maintenance tip on safety valves is to make a safety-valve gag (see sketch, 2.9B). This will prevent maintenance men from putting too much pressure on the spindle and bending it while running a hydro test on a boiler. A 4-in.-long piece of round bar welded to the top of the threaded piece will give sufficient leverage.

F FENNELL, *Westbury, NY*

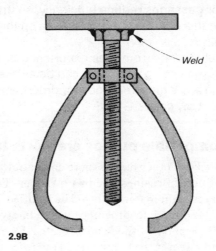

Weld

2.9B

Worthwhile study could last several years

In plants in which I have worked, we found it more economical to check and repair safety and relief valves during our regular annual maintenance of equipment. We also found that it was a good idea to blow the valves by hand each day. This helped to keep them free of dirt and corrosion. A good record of valve maintenance is important, too. If a record shows when repairs were made or a new valve installed, the first leakage should also be recorded. Information on type of chemicals in the vessel and location of the valve will help in determining how long the valves will last on various tanks and equipment. This might take several years, but the result will be worth the effort.

R WINCHESTER, *Pinawa, Man., Canada*

2.10 When should worn pump be scrapped?

Pumps and their maintenance are a big slice of life for us in our chemical operations. We have a large inventory of pumps on process work—charging, recycling, circulation, and transfer. Most of our pumps for corrosive chemicals are made of CF-8, which is a stainless steel alloy, or of high-silicon cast iron. We have a few all-bronze pumps and some all-iron ones, too. Nearly all our pumps are single-stage centrifugals with capacities to about 800 gpm. Some of our cooling-water circulating pumps are much larger, of course.

Our chemical pumps require considerable maintenance. Opened up for inspection and parts replacement, many show wear of the casing interior. Whether this is all corrosion or in part erosion is difficult for me to tell. Whichever it is, I believe that the change in shape and the roughening of the interior passages reduce efficiency, so that the rebuilt pump will cost us more than a new one. I have no data on this and cannot make accurate tests.

Can POWER readers tell me from their experience what are the economic limits to which pump casing corrosion can be allowed to go? And have POWER readers found any simple methods for determining pump efficiency to help me document my case? — BNE

Better, more compatible pumps are worth tryout

Why doesn't BNE look for other pumps more compatible with the fluids he is pumping? The extra cost of casings lined with rubber, Teflon, ceramics, etc, and internal parts of resistant materials could be justified by reduction in parts costs, maintenance time, and lengthening of the interval between repairs. BNE could buy one new pump of this kind, operate it for awhile, and inspect it at regular intervals to discover the pump's ability to resist the chemicals.

A pump requiring excessive maintenance and parts replacement could indicate:

■ Inferior maintenance procedures
■ Substandard parts
■ Poor pump design
■ Use of pump for purpose for which it was not designed

J P GERVAIS, *Medley, Alberta, Canada*

Here is a one-point plan for finding pump efficiency

BNE can measure pump efficiency simply: Determine the capacity and corresponding head at a point close to the pump's guaranteed operating condition, then plot the point on the head-capacity curve and see how the performance compares with the new pump's guarantee. The manufacturer can supply a head-capacity curve if BNE does not have the original. The physical measurements are not difficult. Capacity may be found by timing a change of level in a feed tank or discharge tank. Make sure that the entire change in level is caused by the pump and not partly by leaking or wrongly positioned valves. Portable flowmeters could be used but can be very expensive in chemical service.

Head is the difference between readings of two pressure gages, one on pump discharge and the other on pump suction. The gages should be lab test instruments used only for this purpose and removed when not in use. Remember to calibrate the test gages to assure reliable results. Permanent plant instruments are not good enough for test purposes. The shutoff head, or pressure developed with the discharge valve closed, reflects to some extent the internal condition of the pump. Compare this head with the value at first installation of the pump.

T W BARSTOW, *Redondo Beach, Calif*

Casing shape, yes—surface finish, no

At our refinery, where I am in charge of 1500 pumps, we have been able to cut costs considerably in maintenance and operation, so our philosophy may help BNE. First, casing deterioration such as he described usually results from both erosion and corrosion in localized zones of turbulence. Casing tongues or cutwaters are highly vulnerable, especially when operation is far from the best efficiency point. I have experienced vibration and shaft fracture on double-volute pumps where the tongues of the second volute had been badly damaged. On one pump, the tongue had disappeared completely, destroying radial balance and causing three broken shafts in a year. Casing tongues have to be repaired carefully to avoid the problem. Volute gap between impeller and tongue must be at least 5% of impeller diameter.

Rough casing surfaces do not cause noteworthy efficiency loss. Some years ago, I was associated with tests to evaluate the effect of highly polished casings finished with special paints. Efficiency gain was only 1-2% over that of the roughest casing finish.

Impeller surface finish is another matter. Roughening from corrosion or deposit formation on outside of shrouds or internal passages will drop efficiency and raise power. Reason for the difference is simple. On process-service radial impellers, mean fluid velocity over shrouds is 3 to 4 times that of fluid over casing surface. Skin friction loss is proportional to speed, so for a pump in good condition the impeller shroud loss will be about 10% of total power, while casing friction loss will be about 1-2%. Casing produces a shock loss as the fluid is impelled against the tongue; this is another reason for repairing the tongue carefully to its original form and polishing over the first one or two inches only. BNE can forget about roughness in the casing, provided that there are no weak spots that may break through into the suction passage.

For the important site testing to prevent catastrophic failure, I use a simple method involving knowledge of only process flow, pressures at suction and discharge and line amperage. Motor efficiency and power factor are available from factory data. Total head is discharge pressure minus suction pressure, divided by specific gravity and multiplied by 2.31. Useful hydraulic horsepower equals flow (in lb/min) times total head divided by 33,000. The pump's efficiency equals useful hydraulic horsepower divided by brake horsepower input to the pump. Take readings for three or four process flow rates and plot head, power and efficiency against flow. Comparison with the original test performance will indicate deficiencies and whether a complete breakdown is imminent.

W P HANCOCK, *Caracas, Venezuela*

Coatings could be best remedy

I believe that BNE is attacking the problem from the wrong direction. His single-stage pumps are the clue to a simpler solution than his present one. Some of the numerous coatings available, from polymers to ceramics, should aid in retarding erosion and corrosion in volutes and chambers. One pump now available is built with a circumferentially split casing protected inside by a renewable polymer or rubber liner.

H H Leidolf, *Salem, NJ*

Carbon and nickel alloys work well

If BNE is to document his specific case to determine centrifugal pump efficiency, he will have to plot characteristic curves showing head, efficiency, and brake horsepower as function of discharge. To detect changes in metal thickness caused by erosion or corrosion, the differential transformer is a useful transducer. This particular transducer works equally well on magnetic and nonmagnetic metallic materials.

While stainless steel, monel, ni-resist, ductile iron, and nickel-chrome alloys can provide the necessary resistance against corrosion when pumping a wide variety of liquids. BNE should also consider carbon graphite parts, which can be used as vanes, end-plate bearings, seals, pistons, and cylinder liners. They resist even concentrated hydrofluoric and hydrochloric acids. For abrasive sand-water mixture, tests show that special abrasion-resistant cast iron impellers have four times the life of bronze types.

Nickel iron is a good choice for casing and impeller, with a monel alloy shaft. To make sure that the impeller does not suffer severe cavitation effects on the inner walls, chromium content should be 3.2%. The casing (much less subject to such effects) should have a 2% chromium content for long service. Wear rings of nickel iron with 2% chromium will perform better than the chromium stainless steel type.

R G Pheil, *Racine, Wis*

Is manufacturer's corrosion allowance used up?

Upgrading materials will help BNE to reduce his parts inventory for liquid ends. For instance, spare casings and impellers of stainless might serve for both cast iron and stainless pumps of the same size. Corrosion can be differentiated from erosion by the specific patterns of metal removal. Corrosion will produce random pits and marks over the entire contact surface. Erosion occurs along well-defined line patterns and on high-velocity areas, such as impeller tips. A good indication of need to replace a casing is the exhaustion of the manufacturer's corrosion allowance. Beyond this, pressure developed by the pump may exceed permissible stress. Also, the advice of a manufacturer's representative could be useful.

D L Feray, *Tulsa, Okla*

2.11 Best steps in updating old piping

Our new management team is trying to revitalize a large and not-too-efficiently-run chemical plant. An old piping system is one of the headaches with which we must live. Inadequate capacity in some sections, corrosion and internal deposits, poor line support, leaking valves and joints—these are some of our difficulties. We have been trying to keep up with repairs to the system, which includes lines for steam (300 psig and downward), condensate, water, process chemicals, process gas and compressed air. I have been looking for some way to get ahead in repairs, on our limited budget, and to bring the system into shape without spending additional large sums. What I would like to find is some single repair activity I can work at persistently, which will bring the best overall results for the system.

Can POWER readers suggest any experience-based ideas along this line? Just one activity at a time, however. I am deluged with recommendations to rip out all the piping, replace all the valves, or rebuild all the lines, and I just don't have the money. A simple activity that I can blend in with day-to-day operations is needed.—FLG

Pipe supports are priority when revamping

FLG is not the only engineer facing this sort of problem. It cannot be solved overnight, and I don't believe any one "extra" activity is going to get the plant out of its trouble. Lack of profits over many years could be the real reason why piping is in such bad shape there, and unless the new owners can bring in some new money all FLG's efforts will be wasted.

From what FLG tells us, however, I reason like this:

■ Inadequate capacity of the lines? This is not FLG's business to determine. It's strictly a process question. If the lines should be bigger, the management must decide and then must appropriate the money. The replacement work will come to FLG as a matter of course.

■ Corrosion in the piping? FLG can forget about this as an extra work area, too. Corrosion effects show up by themselves, so they will become part of the normal workload. Also, one of the surest ways to make a minor repair job into a big one is to begin by repairing a small corrosion-caused leak on a piece of line. All too often, more and more defective pipe shows up, so that the entire line must be replaced.

■ Internal deposits? These don't usually harm a line. Where they cut down capacity, the question is again one for management to decide.

■ Leaking valves and joints? It's possible to work out good programs on these items, but I think that, for the time being at least, FLG should start elsewhere.

■ Poor line support? This is the one I single out as most important. It is not a defect that is often written up on work-request forms from the plant, but it is potentially very dangerous. Remember that some of the other troubles FLG is having could come from sagging and badly aligned lines, and also remember that human life is endangered if lines fall or fracture because of support failure. FLG should start a small program on this, making up a few hangers or one support at a time. Most of the installation can be done while the plant is operating, so little overtime will be run up.

R SAMSON, *Houston, Tex*

Get an outside engineer's opinion

FLG's management should hire a professional engineer to survey operations and determine what will make them efficient and profitable. The entire redesign of plant should specify how to replace the old apparatus gradually on the way to the new plant. Replace the danger sources first. Consider the boilers, too, because a reduction in pressure from 300 psig might leave the plant without the necessary pressure.

W JACKSON, *Forest Hills, NY*

Top priority: Go after the pipe supports

With advances in the art, obsolescence, and increased production, a chemical plant changes so much over the years that it has little resemblance to its original concept. If original equipment has been replaced by larger and heavier units, the floor loadings may have been exceeded, with structures overloaded to the point of harm to piping supports and restraints. Inadequately supported piping can throw undue stress on equipment terminals, valves and flanged joints. Result? Leakage and poor drainage. Priority should go to repair of faulty pipe supports and contributing building-structure defects.

FLG should understand that this is only a temporary expedient. A long-range plan for best results requires a complete set of drawings of the piping system. A multibuilding plant needs a plot plan showing outside pipeways. Floor plans of each building should show equipment locations and indicate floor loadings. Main-line isometric drawings to show all branch lines, terminals, valves, supports and restraints, plus standard details for future installation of all traps, control valves and instruments, will help FLG note his repairs and trouble spots.

J BRODSKY, *Newark, NJ*

Sketches and notes come first

Lines most needing repair or replacement should be worked on first. These could be process or utility lines. In any case, the best time to make repairs is when the plant is not working. This could be at night if the plant operates single-shift. If a plant runs on a 24-hr basis, it might be wise to run a parallel system and connect in at the first opportunity.

Before any work is done, however, FLG should study all the defects and sketch out the lines, with notes on what should be done. Possible tasks are: replace entire lines, replace pipefittings and gaskets, or maybe just replace gaskets. And remember to get all material assembled at a convenient nearby location first.

E Kaknics, *Philadelphia, Penn*

Stop the leaks, then replace leaky valves

From the standpoint of operating-cost savings, the first activity FLG should approach is stopping of line-joint leakage. Next, he should replace leaking valves. Replacement of poor line support will reduce strain damage to the system and prevent sudden and costly piping failures. And another starting point for the modest effort that FLG can afford is replacement of inadequate and fouled lines to improve plant operation. One step that FLG did not mention is improvement of insulation. No matter which one he chooses, however, he should stay with it and not jump to other activities.

B B Brown, *Hagerstown, Md*

Repair your valves and traps

Valve and trap repair is still the best shop activity for short lulls between outside work. If the repairs are to be cost-effective, however, the valves and traps must be of types that can be repaired without excessively expensive parts and without high labor hours. In recent years, we have seen some valve and trap makers turning out products that can't be repaired, so FLG must avoid these, unless the price is farther below that of a repairable unit than I have seen.

Best results will be on medium-sized units, such as valves from 1¼-in. to 12-in. size. Smaller valves should be saved up and repaired in batches. For bigger valves, FLG could consider outside repair services. Standardizing on reputable manufacturer's repairable models is always wise. Any price premium is recovered on the first repair. Remember, too, that the cost of the repair is not as evident to management as the purchase of a new valve would be. Labor for installation, removal, and replacement is the big item today, especially with small-size fittings. In our experience, time between overhauls for repairable equipment is longer than the life of the nonrepairable kind, because once a maker starts designing for throwaway, he takes many shortcuts that are not possible on equipment capable of repair. FLG should also test each repaired unit before putting it in the stockroom.

T W DeLong, *Peoria, Ill*

Marking pipelines pays quick dividends

Before FLG does anything else, he should send someone out to mark every line in the plant with the contents and direction of flow. We lost considerable time and maintenance money before we woke up to this simple expedient, which is quite common in commercial and building work but unfortunately often lacking in industry and utility plants. Color coding is good, but the main objective is to identify each pipe by some marking that will let a fitter find it quickly without tracing out drawings and walking around the boilerhouse to find where the line is coming from or going to. Several companies make wrap-around or stick-on markers. These are worthwhile helps in this simple, cheap, and highly cost-effective plan for improving piping systems maintenance.

T WELD, *Hollywood, Fla*

3 Fuels and Fuel Handling

3.1 Ways to eliminate coal-pile dust

Our plant has been increasing its stockpiling of coal during the past three years. We had not had extensive land area for this, but by taking over part of an outside storage yard, we obtained a long strip about 100 ft wide and 1000 ft long. We have built up a pile about 80 ft wide and the full length of the area, using a small bulldozer, which compacts the pile and reclaims the coal to the conveyor when needed. We probably have about 30,000 tons in the pile now.

Our main trouble is with dust from the pile blowing out into local yards and houses. Prevailing winds are across the pile and can often exceed 20 mph for many hours. We have tried chemicals on the pile but they were only marginally effective and were expensive. We try to use coal from all along the pile, to prevent any of it from standing and weathering too long, and chemical loss is therefore probably more than for some other plants. Planting of trees or building of a screen has also been suggested. What experience have POWER readers had with measures, such as water spray, to suppress dust loss from coal piles? Does the shape of the pile affect the loss? Would a rubber-tired vehicle make for fewer fines than our present bulldozer does?—JD

Manage live- and dead-storage areas

There are no surefire solutions to keeping fugitive-dust emissions in check, particularly in high winds. Many factors help to create a fugitive-dust problem, including:

■ Coal properties: chemical makeup, friability, size distribution, moisture content.

■ Weather conditions: rainfall, humidity, wind speed.

■ Stacking/reclaiming methods: truck, conveyor chute, bulldozer.

Our experience with coal-fired powerplants in India indicates that occasional water spraying is the only feasible method of suppressing coal dust in outdoor piles. Permanently mounted water sprinklers located in troublesome areas such as unloading chutes have been successful in some applications. We found that water additives to help in fines agglomeration work marginally at best.

Dust control is best achieved by adopting well-coordinated inventory measures for material, and by selectively segregating live- and dead-storage zones. Dead storage should have suitable permanent or semipermanent surface protection such as grass growth, coal tar, or pitch, and should be used only in an emergency. Reclamation should be restricted to a relatively small portion of the total pile, with constant attention paid to dust control by water dousing. We do not think that screens provide a sound engineering solution to controlling fugitive dust. We also reject the belief that rubber-tired vehicles create less fines than bulldozers.

B K MITRA, *Philadelphia, Penn*

Water and compact it, or bury it

Our plant maintains a 2.1-million-ton coal pile with prevailing winds ranging up to 60 mph for several days at a time. We have found that chemicals are not a very effective dust suppressant. We use water instead, mainly to assist in obtaining optimum compaction of the coal.

The best way to limit fugitive-dust emissions is to make part of the pile a dead-storage area. We water this section, compact it with a vibratory roller, and leave it alone. Of course, this approach assumes that oxidation is not a significant problem. We have determined through fuel analysis that when our pile is compacted to an average of 70-80 lb/ft^3, heat content of the coal won't seriously degrade until after five years of storage. After this period, we know that it's time to start rolling over the area. Regarding dead-storage areas: Count on having high coal-dust blowoff after initial buildup but a fast decrease in emissions thereafter. Also, if you are in a high freeze/thaw area, you should expect to see a slight increase in emissions after each freeze and thaw-out period.

Another solution is to put the coal pile in a subsurface storage area—if the water table permits—so that the top of the pile is below grade and out of harm's way. Consider also creating a large dead-storage area and a small active or live-storage area to provide maximum shielding of live areas from the elements (see sketch). This will reduce, but not eliminate, your fugitive-dust problem. Only protective covering will contain emissions completely. In our experience, there appears to be no difference in fines production between rubber-tired and track-type bulldozers.

G LEARY, *Page, Ariz*

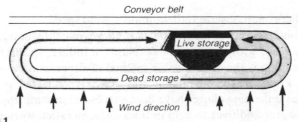

3.1

The downside of spraying with water

If you spray stored coal with water, it may crumble and become more susceptible to spontaneous combustion, particularly if the pyrites content is high. When properly compacted, a rectangular pile with sloped sides and a crowned top to aid in drainage will deter wind losses.

If outdoor storage won't be used for relatively long periods, consider covering the pile with an airtight asphalt or road-tar seal. Surfaces exposed directly to prevailing winds may require greater thickness. Care must be taken to prevent rupture of the seal by moving vehicles, picks, and shovels. Some type of provision for thermal expansion and contraction of the pile should also be considered. In some cases, construction of a concrete enclosure for a pile may be justified.

H B WAYNE, *Jamaica, NY*

Don't discount chemicals completely

JD said that he had tried chemicals without notable success, but he may not have realized that there are several types of chemicals available. The newer ones are reported more effective in fugitive-dust control. According to reports from the manufacturer, these polymers form a relatively long-lasting bond layer on the surface of the pile. The surface is permeable, so there is no alteration in water-runoff characteristics. The bound layer also reportedly resists breakage when the pile settles. If these claims are true, then maybe JD should give chemicals another try.

W F FLOWERS, *Oak Brook, Ill*

Use chemical sprays more effectively

Surface-treatment techniques may be the only hope for low-grade coals—such as low-rank subbituminous—which contain an excessive amount of fines and degrade under the effects of weathering. Several organic polymers and a petroleum resin/water emulsion are the most commonly used sprays. Most are designed to be compatible with the environment, resist degradation from bacteria, oxidation, and sunlight, and will not interfere with the quality or subsequent use of the coal.

The success of chemical crusting as dust controller, however, depends on several factors, including:

■ **Pile preparation.** The pile should be compacted to ensure a competent base for the crust. Sides should have at least a 2:1 slope both to allow for thorough and convenient compaction and to keep the crust from cracking and separating. Piles should be shaped to remove sharp edges or ridges that would be prone to wind erosion, and to avoid partial vacuums or eddy currents that would lift the crust.

■ **Application method:** The only equipment needed here is a tank truck with a pump that can develop enough pressure to spray all portions of the stockpile.

Some plants also use crusting agents to obtain benefits other than dust control. By using a latex polymer compound, for example, we've been able to seal off the pile from the effects of weather. With this type of treatment, the pile actually loses moisture. Reason: A greenhouse effect is created by the covering, which reduces the chance of spontaneous combustion and weathering.

S B KIERNAN, *Providence, RI*

3.2 Specs and tests to get better coal

We are a medium-size industrial firm generating process steam at four locations in the Midwest and Southeast. We have oil-fired, pulverized-coal-fired, and smaller traveling-grate boilers. With the upswing in interest in industrial coal-burning, we find that the market in the coal we have been burning has been tightening. In addition, we plan a complete switch to coal for all plants. It looks to us like the wave of the future for industry. Because of this, we are looking closely at improved specifications for coal.

We would like to set up specs on coal, and also standards for tests. The question of penalties and premiums has come up, too. We realize that some of the practices in utilities could also help us. What has been the experience of POWER readers on the need for standards and specs for such characteristics as sulfur, ash, fusion temperatures, grindability, and delivery method? How detailed should analysis be? How close should our tolerances be, and what kind of lab and equipment should we have? Are there reliable methods for weighing various factors to compare coals?—JSC

ASTM standards are recommended

A number of standards of the American Society for Testing and Materials (ASTM) are recommended for determining coal properties. Here are some of them:

ASTM no.	Test analysis
D 197	Fineness test (pulverized coal)
D 409	Coal grindability
D 440	Drop shatter test
D 441	Friability
D 720	Free swelling index
D 1857	Ash fusibility
D 2015	Gross calorific value
D 3173	Moisture content
D 3174	Ash content
D 3175	Volatile matter

D 3177   Total sulfur
D 3178   Carbon and hydrogen contents
D 3179   Nitrogen content

All these standards give the procedures and required facilities for performing the tests. For a reliable analysis, get several coal samples, so that average values will be represented. Another ASTM standard, D 2234, on gaseous fuels, coal, and coke, has reference to this in its part 26. Taking samples of coal while the coal is in motion is the preferred way—this means during loading and unloading on conveyors and other handling facilities. Sample either continuously or intermittently. For multi-plant coal contracts, here are recommended data:

Mine location, name of seam, coal size, guaranteed ultimate analysis, ash-softening temperature, heating value (to include minimum value), shipping point, type of transportation (whether marine, rail or truck), and name of the transport contractor. Often coal properties vary between long-time intermittent deliveries, meaning six months or longer. A bonus/penalty arrangement can compensate the purchaser for the failure of the seller to satisfy contract minimum values. Typical variables include increased operating cost resulting from handling and disposing of greater amounts of ash, and lower heating value, which calls for greater fuel consumption. Also, higher sulfur, nitrogen, and pyrites content means increased maintenance and possibly environmental problems.

Unit cost for a long-term contract is usually less than for short periods. Nevertheless, the latter type may be more economical when coal prices decline in market fluctuations. Because many coal contracts extend for relatively long periods, clauses are usually included to compensate for the seller's anticipated increase in labor costs. Finally, if plant location and coal-handling facilities allow delivery by marine, rail, and truck shipment, flexibility is better, with less chance for supply interruption from factors like inclement weather or strikes.

H B Wayne, *Jamaica, NY*

Geography is influential with coal specs

Swinging over to coal is an involved change, requiring considerable justification, planning, and preparation. JSC certainly needs coal specifications, which will have to vary according to the geographical location of the plants if consistent performance is sought in all the plants. This is true unless he is prepared to pay a penalty for freight charges so that he will get coal to his precise requirements. Going over the characteristics in order:

■ **Sulfur content** of coal varies from as much as 5% to as little as 0.3%. Some of the central-state mines produce the higher levels. Because control of sulfur emissions is a strict requirement in most places, JSC should evaluate cost of low-sulfur coal vs cost of emission suppression.

■ **Ash content** generally varies from 4% to 15%, and the ash itself can be a problem, depending on firing methods and furnace temperatures.

■ **Ash-softening temperatures** need to be specified according to the ash content of the fuel. Remember that higher ash content usually calls for a higher softening-temperature specification.

■ **Grindability** is one characteristic that varies from source to source and depends on the mechanical structure of the coal. A lab procedure relying on a specialized test machine determines grindability.

■ **Delivery methods** ought to be the option of the customer. They depend on what handling facilities the customer has at his sites. Specifying a delivered bid price relieves the buyer of this concern if his handling abilities are very flexible.

Although there is no set of simplified answers to JSC's questions that can be written briefly, he should remember that quantity of coal burned is an important factor determining the amount that can be spent on quality control. Up to some point, an independent lab is the more economical control and testing means. For larger amounts of coal, an in-house lab is justifiable. ASTM and US Bureau of Mines publications on coal specs, testing, and characteristic qualities are highly recommended in any case.

B B Brown, *Hagerstown, Md*

Start with boiler review

First step in specing out for coal is to review the boilers that are going to burn the coal. This means getting out the old drawings and calling up the manufacturer, if he is still in existence. In many cases, too, there are plants in your area that have similar boilers and can give you some valuable clues on what not to burn in the way of bargain-basement coal.

How hard JSC forces his boilers is another question that affects quality of coal bought. If there are no big load swings, and if the rating is conservative, chances are that a poorer quality of coal can be burned. Another thing to look out for is whether any fuel-saving equipment, such as economizers, has been added to the original boiler. If so, then the flue-gas temperatures are sure to be down, and the limit on fuel sulfur and perhaps other constituents, too, is going to be lower.

We ran into this problem ourselves a short time ago and found that we needed to cut out two of our long-time suppliers because they were unable to give us the low-sulfur coal we needed, at a reasonable price. This whole question is apart from the environmental one, of course, which up to now has not been really serious for the smaller boilers.

G Atchley, *Toledo, Ohio*

Make each plant responsible

JSC may not be well advised to force uniform standards and procedures on all the boiler houses in the system. Unless he is prepared to exercise close supervision, which is not easy, the best way to keep fuel costs down and boiler performance near peak levels might be to let the separate plants buy their own fuel for a while, with a few simple checks on costs per million Btu and total cost.

There are many details of fuel procurement and burning that differ from plant to plant, so that forcing a plant to take a coal it does not want can be counterproductive. On the other hand, if an individual plant competes with its own past record, the plant will have an incentive to improve and will suggest ways to reduce overall cost not only in fuel but also in indirect expenses, such as ash handling. Once JSC knows what the plants are comfortable with, he can begin to tighten up specs and institute uniform reporting procedures and centralized buying.

R W Masaro, *Pittsburgh, Penn*

3.3 Unexpected costs in burning coal

Planning for a change to coal burning is occupying our attention now. We are considering new stoker-fired boilers in the 200,000-lb/hr class. We have never had any experience with coal, only oil and gas, so we are working with a small but reliable consulting firm and also are talking with as many equipment manufacturers as possible.

In estimating costs of operation, we are not on as firm ground as we are for equipment. Although fuel and operating-labor estimates look fairly solid, we think there may be occasional heavy expenses that we should know about and plan for, if we are to give our management a fair picture of what is ahead during the next few years.

We know there have been many horror stories about what has happened to plants burning coal—things like changes in the coal, long runs of frozen coal, coal-pile fires, and work that had to be done on coal piles to stop pollution. Perhaps some planning now can lessen the chance of unpleasant surprises in a few years' time. Can Power readers tell us what has been their worst-case situation in burning coal? Where should we put our major planning emphasis? Should we sink money into a big coal stockpile, or is that merely a delusion of safety? How flexible should our ash-handling be? And how much of our repair work should we staff up for?—CNV

Coal trucks trigger resident complaints

Complaints and friction with local residents and authorities have been our worst problems in coal burning. From our experience, CNV should try to get his coal shipped in by rail rather than truck. It seems that no one cares about coal that falls off railroad cars and lies along the track siding, but everyone in town starts to protest when a few pounds of fines and lumps slip off a truck onto a town road.

With ash disposal, trucking is probably the only way to take out the material, so any complaints there have to be lived with—following best efforts to clean up whatever the reason is for the aggravation. Totally enclosed trucks are out of the question. Covers and some wetting with water, or perhaps one of the chemical treatments, are about all that can be done. The contractor who takes away the ash is the low bidder on the purchase request, and he has to be watched closely to make sure that he is not overfilling the trucks and letting part of the load spill along the way to wherever he heads.

D K HEATON, *Cincinnati, Ohio*

Top-management help needed

To make a conversion like the one CNV is talking about effective, there must be some executive in the company who takes a real interest in it. Remember that although coal salesmen like to sell coal, no one likes to burn it. At the slightest hint of an easing in the oil-and-gas supply picture, there will be a determined effort in many companies to go back to the easy fuel and put the coal-burning ventures on hold. But that is the road to big trouble in a few years, when the fuel picture changes again and the oil- and gas-burning plants are again hard-pressed to find fuel.

In a multiplant outfit like mine, it is possible to make some compromises. The most suitable candidates are forced to burn coal first. Then, when they have run up some experience and found out that coal burning is possible, other plants can be gradually put on coal. One general rule to remember in switching from oil/gas to coal is that everything is bigger—boilers, fuel-handling area, ash-disposal area, and repair shop. Also, maintenance problems and emphasis will change. CNV will need more spares on hand for such equipment as stokers, elevators, belts, and sootblowers. It is better to have too much on hand than too little. The design of most coal-burning equipment doesn't change quickly, so what is bought today will be effective for a long time.

H T BUTLER, *Indianapolis, Ind*

Keep plenty of coal on hand for continuity

A large dead-storage coal volume offers greater assurance of operating continuity if shipments should be interrupted. Adverse effects of labor strikes in coal mining and transport, and of inclement weather, would be cut or eliminated. Having a large coal inventory gives you a better bargaining position when new supply contracts come up for discussion. As for operating problems, arching and ratholing of coal in bunkers and silos need attention. Coal moisture decreases coal-handling capacity, beside corroding equipment and bunkers. Sulfur may lead to heating and fire inside enclosed storage facilities. An inert-gas installation, with CO_2 or N_2, is common as a preventive measure.

Stokers call for close control of coal size. Operators must prevent segregation and keep a uniform fuel-bed density. Soot-blowing requirements for coal-fired units usually exceed those for oil-burning installations. Something that is not a problem with oil- and gas-fired units (but can be with coal setups) is over-heating of stoker components because of frequent and fast-changing overload firing rates. This can mean high maintenance costs and parts replacement. Daily inspection of accessible elements of the stoker is recommended.

Selection of dust-collector type depends on coal properties, stoker type, and firing modes. Collection efficiency and maintenance needs also depend on these factors. For 200,000-lb/hr steam capacity, spreader stokers will probably be the choice. Amount of flyash, which ranges between 12% and 60% of total combustion ash, depends on coal type and ash content. Baghouses frequently are the best type of dust collector for the service described. Even though first cost of negative-pressure systems usually exceeds that for positive-pressure, fan maintenance is less.

H B WAYNE, *Jamaica, NY*

Ash-disposal costs are steeper

Ash and sludge disposal will be a big unknown for CNV, no matter how detailed and permanent the solutions look right now. With constant change and tightening of regulations, a disposal site that would have been good for many years can suddenly become unavailable, or available only at a very high add-on cost for improvements. For ash disposal, CNV should remain alert at all times for opportunities to get rid of some ash cheaply. Every now and then, a builder or dump operator can take some for temporary roads or certain types of fill, and the amount taken can reduce disposal costs significantly. A smaller plant, with moderate amounts of ash, is less exposed to public criticism than is a big utility station, and flexibility in working with the ash problem should be put to good use.

K I BRIERMAN, *Pittsburgh, Penn*

Be ready for unloading in winter weather

If CNV is in the northern part of the US and expects to get coal in by rail, freezing will be his biggest headache. This has been our experience. We burn some two to two-and-a-half cars a day during winter, but in time of cold weather, deliveries are not closely predictable, so we may have to depend on a small stockpile for several weeks at a time. Big stockpiles are hard to sell to industrial management, who prefer to rely on skill in buying coal around the country or on a switch to oil for a short time in case of strikes.

Thawing sheds and other expensive equipment are probably not possible for a user in the 200,000-to-600,000-lb/hr class, where CNV appears to fit. Shakers, heaters, and burner setups are the chief means for small-rail-shipment receivers to get frozen coal out of the cars. It's all very well to talk of chemical treatment of coal before shipment, but only big utilities and industrial customers can enforce such treatment. Small industrial buyers are often at the mercy of suppliers who push untreated or half-treated coal on them. When a string of frozen cars comes in, the alternatives are working on the cars or trying to get frozen coal out of the stockpile. Either alternative is a horror, and CNV should be ready for them, with a gang that can perform miracles at 30 below.

R L DANNING, *Milwaukee, Wis*

3.4 Cutting coal-handling maintenance

We started up our first coal-burning plant 18 months ago. Our many years with oil and gas fuels had not prepared us for the trying experiences we have had. Our pulverized-coal setup, fed with Illinois coal, can send 600,000 lb/hr of 450-psig steam to process when everything is in order. We have had several outages traceable to precipitators, induced-draft fan, and water treatment, but what I am most interested in just now is the high maintenance and downtime on coal-handling equipment. Elevators, conveyors, crushers, vibrating feeders, tripper conveyor, and fabric filters at coal-transfer points are prime expense causes. Abrasion and belt wear seem high, too.

For example, we have to replace at least one conveyor-roller bearing every week, even though the bearings were supposed to go at least one year without attention. Corrosion is another source of expense for structural repair and painting. We calculate our maintenance costs for coal-handling equipment to be 63¢/ton of coal (1980$), much higher in relation to energy than for our gas and oil plants. What has been the experience of Power readers with coal-handling system costs? What are effective solutions to the specific problems that we and others have? How can we draw up better specs for future coal-handling installations? —TG

TG's costs are in line

It is axiomatic that fuel-prep cost is certain to be higher with coal than with oil or gas, and equally that maintenance expense will be higher. Bottom line in the whole situation, though, is the price per million Btu, kWh, or other output unit. References in my files show total maintenance costs for coal handling and processing in electric-generating plants using Illinois coal to range from 1.5 to 2.75% of the cost of fuel. If TG's fuel cost is $30/ton, he is in the center of this range. Even at $40/ton, his maintenance/ton would not be excessive.

To reduce the outages, a preventive maintenance program that will anticipate breakdown is a good idea. Immediate relief should result; that is, within a few months. This should not be confused with replacement by more reliable, longer-lived systems, of course. The start of the PM program should include frequent inspections of all the equipment for indication of failure. Noisy bearings, small breaks in belting, and vibration in drives are examples.

Simultaneous record keeping for every breakdown or failure of any part of the system is also necessary. Establish and follow a regular housekeeping, cleanup, lubrication, adjustment, filter-cloth replacement, and induced-draft-fan cleaning program. There is no doubt that PM is front-end-loaded in cost, but as TG's experience develops, the cost of inspection, etc, will drop and operating reliability will pay for the expense.

As time passes, a predictive-maintenance program will succeed the preventive-maintenance program, and then attention goes to equipment adjustment, lubrication, repairs, overhaul, and replacement on a firm schedule based on maintenance records. The predictive-maintenance effort will substantially reduce outages from unforeseen breakdowns, and emergency breakdowns will drop to a more tolerable level. William Staniar's "Plant Engineering Handbook" is an invaluable source of PM guidance.

Once they have gone through the preventive and predictive maintenance exercises, TG's group will be an expert source of guidance for specification preparation on succeeding plant systems. The group will know exactly where weaknesses are and what to install for longer service life.

B B BROWN, *Hagerstown, Md*

Get out and get under

I wonder how long is the interval over which TG is figuring his maintenance costs. A year is too short an interval for the type of equipment concerned here. One large repair job can distort the results badly over a single year and can trigger hasty decisions leading to useless expense. TG should wait a little longer until he can compare at least two sets of seasons, say. Winter maintenance will be much more than for summer.

A good start might be a careful study of just what the maintenance and especially the preventive-maintenance people are doing. Are repairs being made merely to bring the equipment back to its original state, or does someone look into each major breakdown to find out if change in components, adjustments, or lubrication is going to be cost-effective?

Is there any point at which a minor addition or change will improve matters enough to prevent an unexpected breakdown? A deflector or guard over a line of roll bearings at a loading point, or a slightly reduced feed rate at a point where pileups or clogging have occurred, are examples of what I mean. These opportunities don't announce themselves. You have to go out and look for them. They should be found during a proper shakedown and trial run, but sometimes plant management is in too much of a hurry to get into full production and doesn't let the resident or plant engineers go searching for vital opportunities.

J E MANNING, *Baltimore, Md*

Six ways to upgrade the system

Here are some tips for TG to help reduce his problem:

■ Abnormal conveyor-belt wear has several causes. Excessive impact of coal from feed chute onto belt surface is the result of high loading velocity. The chute discharge flow should be at conveyed-material speed, in the same direction as belt travel, and centered on the belt. Impact idlers could help.

■ Get rid of excessive belt sag. The acceptable amount depends on idler spacing, belt tension, and belt weight of a length between idlers. Reducing idler spacing or increasing belt tension will help. Don't overstress the belt and drive assembly by excessive belt tension, however.

■ Prevent coal buildup on belts. Because coal particles are abrasive, an accumulation can penetrate the carcass. Scrapers help, and installation of a grizzly or screen in the feed chute allows fines to drop on to the belt before the lumps do. The fines cushion the belt.

■ Keep the belt aligned. If misalignment is excessive, the belt may rub against surfaces. Bearings can then be overloaded. Self-aligning idlers help here.

■ Bearing-assembly sealing must prevent rain, snow, condensation, and coal dust from entering.

■ Corrosion of steel structures in certain areas can be severe. Sulfur compounds in the coal are a cause. A self-curing inorganic zinc primer, applied on a scale-free near-white metal surface, protects. Two topcoats of an organic sealant on a completely dry primer will limit corrosion.

H B WAYNE, *Jamaica, NY*

Special consultant can help

Most industrial plants are not in a position to draw up their own specs for one-in-a-lifetime purchases, such as coal-handling equipment. Best solution is to rely on a competent consultant in this field. He may not be the one on the main steam plant, either, because conveying and handling equipment is a specialty that not every consulting firm or architect/engineer has, particularly for the smaller industrial plant.

This does not mean that TG should not collect information on what equipment is working well or poorly for him now. Get all the data possible on how the equipment has performed, and be ready to show this to the consultant, manufacturers, and contractors when the next large project comes along.

K F BLASKOWITZ, *St. Louis, Mo*

Suppress coal dust first

Getting used to coal as a fuel is apparently the difficulty. Specifically, conveyor bearings should last longer than a year, if they are not overloaded and if they get proper lubrication. Corrosion comes from excessive airborne dust in a moist atmosphere, which causes a chemical reaction. TG could consider dust-suppressing equipment or chemicals.   C R KLEIN, *Wauwatosa, Wis*

Beware of hasty lube change

It sounds to me as if TG needs an experienced maintenance chief to supervise work on the coal-handling system. Merely sending maintenance men out on breakdowns does not do the job. The men must be shown what to look for beyond the immediate repair.

Another pointer: On lubricants, don't listen to every story of a magic remedy for bearing problems. If TG is applying a good general-purpose lubricant on his system, there is no gain to be made by switching back and forth on special-purpose lubricants.

E M CARROLL, *Louisville, Ky*

3.5 Will fuel oil deteriorate in tanks?

Our plant fires natural gas as its main fuel. If we wish to maintain our present rate, however, our supplier requires that we have a standby fuel system, and that it be available for service on short notice. To serve the standby fuel system, we keep 60,000 gal of No. 4 fuel oil in underground storage tanks. This fuel oil is now three years old, and we may not need it for several years. We wonder if this fuel will be usable when we need it. We have heard many stories of fuel-oil deterioration in storage, but are unable to find anyone who can assert authoritatively what we may expect. We don't have storage-tank heaters.

Perhaps a general discussion of long-term-storage problems with all grades of oil would help both industrial plants and utilities. There must be a cheaper way of solving the question than by burning and replacing part of the oil every so often. What has been the experience of POWER readers with the effects of water on stored oil? Can rapid temperature change create problems? What about bacterial growth, which has sometimes occurred in lighter fuels?—PCM

A four-step solution to storage problems

Our experience with No. 2 oil may help PCM. We keep 200,000 gal of standby No. 2 in 20 underground tanks. We have found that there *are* problems with long-term storage, as samples from several tank bottoms (nos. 14, 15, 16, 18) show. Tank 14 would cause burner tips to plug. Tanks 16 and 18 seem to have only a water problem. Tank 15 appears to present no problems. Our four-step solution to storage problems:

- Sample all tanks to determine oil condition.
- Pump off any water.
- Filter any tank whose oil would plug burner tips.
- Apply suitable oil treatment.

The age of the oil in each tank is not known, but some of it is over six years old. All tanks have the same weather conditions in southwestern Michigan. Because only two tanks have had to be filtered, however, it would seem that age and temperature are not the main factors in algae growth.

G METZ, *Ann Arbor, Mich*

Oxidation and breathing are keys

Fuel-oil degradation comes from trace metallic elements in the fuel, and from breathing of the storage tank. The trace elements can catalyze oxidation reactions that form insoluble sludge and sediment, especially at elevated temperatures. Oxidation usually darkens the fuel oil. An accelerated fuel-oil stability test indicates the potential for oxidation.

Breathing causes exposure to air and moisture. Moisture can bring bacterial growth and stratification in the tank. Bacterial count and water content are good indicators of the extent of this problem. There is a specific chemical treatment for four problem areas stemming from the two mechanisms above. A stabilizer is for sludge formation, a dispersant fights water contamination, an inhibitor works against corrosion, and a biocide will combat biological growth.

G BORSINGER, *Whippany, NJ*

Tests are questionable to some

No. 4 fuel oil is a mixture of distillate and residual oils. Origin of the distillates will affect long-term storage stability of the oil. References in the literature report that distillates from straight-run fractions give a more stable No. 4 fuel oil than do fractions from catalytic-cracking processes. Irrespective of fuel source, PCM could consider these ideas:

■ Avoid contact with air. Oxygen will form gum and sediment harmful to combustion. A nitrogen blanket on the storage tank is a possibility.

■ Stability will depend on storage temperature, too. Fuels stored in moderate climates will deteriorate less than those in warm climates.

■ ASTM D-2274 test *may* be used to predict long-term fuel stability. PCM is right, however—data in the literature are often contradictory, and several authors even think that some of the ASTM predictive tests are inadequate or not applicable.

G C SHAH, *Houston, Tex*

Corrosion is not a big problem

PCM's problems will depend on the as-purchased quality of the fuel oil and on storage conditions. Amounts and kinds of impurities will vary from batch to batch, and in-storage factors could include moisture condensed from the air and degradation by oxidation and polymerization processes. Temperature changes should be minor, because storage is underground. Corrosion from separated water that has absorbed acidic contaminants in the oil should therefore be minor.

Oxidation and polymerization will increase with time. Oxidation of unsaturated hydrocarbons in the oil yields organic acids, ketones, aldehydes, and esters, causing corrosion and sludge. Polymerization forms waxes and tars that foul and plug burner tips.

In order of likelihood, No. 4 oil problems will be:

■ Sludge formation through oxidation.
■ Waxes and tars from polymerization.
■ Cold-end corrosion by heavy metals that catalyze the oxidation of sulfur.
■ Corrosion by separation of acidic water layers.
■ Catastrophic corrosion by vanadium.

Chemical additives are available to counteract each of these.

H T BUDKE, *North Kansas City, Mo*

Keeping tanks full will be a plus

As a result of breathing during changes in atmospheric pressure, water will get into oil tanks used for dead storage. The quantity of water will be small, unlike the case with tanks that are filled and emptied regularly so that outside air replaces the removed oil. Over several years, however, water may accumulate and sink to the tank bottom, where attack will cause rust or other corrosion. The water will not seriously affect the oil quality, because oil and water are immiscible under static storage conditions.

Oil will be affected by oxidation from the oxygen in the air, but, if the tanks are full, the area of exposed surface is small. In addition, the top space of the tank will be heavily loaded with petroleum vapors, with a high partial pressure.

B B BROWN, *Hagerstown, Md*

Microbial growth plays a part, too

Microbial growth occurs in most fuel oils, from No. 1 to No. 6, in both surface and underground tanks and is a factor in fuel stability. Microorganisms degrade hydrocarbons and cause emulsions. Such byproducts as organic acids, hydrogen sulfide, and slimy polymers can corrode metal, plug filters, and make water/oil interfaces ragged.

Temperature, pH, and water are important as environmental conditions in a storage tank. Water should be kept to the minimum, because it is essential for microbial growth. It has a dual role of carrying nutrients to microorganisms and removing their toxic byproducts. Operators sometimes keep a continuous water bottom in large storage tanks, however, to reduce localized corrosion. The continuous water bottom evenly distributes acids and hydrogen sulfide, corrosive byproducts of microorganisms, so there is less likelihood of small, highly corrosive water pockets to cause isolated corrosion.

Temperature is important for microbial growth. Optimum span is 60-120F, but the microorganisms can tolerate fluctuations from below freezing to boiling. Microbial growths in tank bottoms are often called green slime or algae. The slime is a matrix of microbial filaments and complex carbohydrate polymers. The green color is from a pigment, and is not algae. Most algae require light energy to grow, and this is not present in most closed tanks.

Slimes, filamentous fungal growth, rod-shaped bacteria, and yeast have all been found in No. 4 fuel oil of utilities, at the water/oil interface. Biocides are available for treatment.

D G Horstmann, *St. Louis, Mo*

3.6 Can diesel oil stop coal-pile fires?

We are one of many large industrial plants that are again firing coal, after years of oil burning. We have had to relearn much of what our predecessors knew but never wrote down, and we are also burning much poorer grades of coal than they ever did. We have had good support from our vendors in reconversion to coal, however. We needed considerable new equipment and replacement parts, too, because of attrition, theft, etc. We will undoubtedly be facing storage-pile fires in the future, and this is a subject that we would like to get some advance information on. We have taken the basic precautions in compacting the coal, dozing it, and keeping the pile sides at a steep angle. We are also making periodic temperature checks along the pile.

The best course of action to take if fire should occur is something we would like to hear about from Power readers. Have there been any changes or new ideas in fighting storage-pile fires? Recently we heard about the possibility of spraying diesel fuel on the pile top as a firefighting method. We would like to know more about this. Does it smother the fire? Is there danger that the oil will seep into a loose pile and feed the fire? Also, can indiscriminate storage of different grades or types of coal promote fires, even if the pile is well-formed?—MFK

Heavy oil can effectively smother the fire

Diesel fuel can be applied to slow the oxidation if the coal is not too hot. The oil will effectively smother the coal and prevent oxygen transfer from free air in the pile. Road-tar oil or even asphalt over the smoldering area is possible. None of these will effectively feed the fire unless there is a blaze. Carbon dioxide can put out the fire. Run a pipe down through the pile and put in the gas to

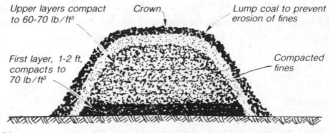

Upper layers compact to 60-70 lb/ft³   Crown   Lump coal to prevent erosion of fines

First layer, 1-2 ft, compacts to 70 lb/ft³   Compacted fines

3.6A

smother the fire. Blocks of dry ice are effective, too. The gas, heavier than air, will drop through the pile and displace the air.

Best approach is to prevent fires before they occur. Don't form a pile over heat sources, such as process lines, drain vents, steam lines, or hot-water lines. Select firm ground without loose fill, ashes, or cinders, and grade and compact it to give proper water drainage. Then store bituminous coal as shown in the sketch (3.6A). For lower-grade coals, make the layers only a foot thick before compacting. Seal the pile with slack-size coal, not lump coal, because lump sizes of lower ranked coals deteriorate quickly.

Freshly crushed coal is more prone to fires, and smoldering warm spots are to be expected. Probe the pile daily, and if the metal rod is too hot to hold in the bare hand, take action. Mixing different grades of coal should give MFK no problem if he takes the basic precautions for the lowest grade coal all through the pile.

R D Hedding, *Newark, Ohio*

Never spray diesel oil on the coal pile

NO—diesel oil cannot be sprayed on any coal pile, especially in event of a fire. If diesel oil is sprayed on any coal, this coal should be burned at once. Accidental spilling of diesel fuel on coal is dangerous and must be avoided.

In MFK's case, obviously diesel oil was mistaken for mineral oil. Under very limited conditions only, high-flash-point mineral oil, but never diesel oil, may be used for dust suppression. In limited indoor or undercover coal storage, this oil is sprayed on coal to protect against spontaneous ignition of coal dust. Even with the highest flash point mineral oil available, however, the total flammability of the coal pile may be increased. Any value in the treatment with mineral oil is therefore only in dust suppression.

Spraying diesel oil on a burning pile will result in a violent holocaust. Diesel oil will ignite, much as it does on reaching the proper combustion temperature in an engine, as it drains down into the coal pile. This is the reason why you should burn coal at once if diesel oil has been spilled on it.

Dozing and compacting coal, as well as checking temperatures, is about the only defense. I have had success with a cryogenic lance for firefighting. Indiscriminate storage of various coal varieties and the mixing of wood and wood products with coal are other dangers that will cause ignition conditions where the lower-ignition-temperature material is. What I have said here is in addition to the basic precautions, given in such books as the "Fire prevention handbook," 14th edition, which is valuable for many operating conditions.

C R Klein, *Wauwatosa, Wis*

Oil, like water, can cool coal

Spraying diesel fuel can provide a more readily ignitable substance for the heat to act upon. If the weather is very cold, however, or if no water is available, oil could cool the hot pile enough to delay or curb the temperature rise. If the decision is for diesel oil, then control the conditions carefully and be prepared to act if something goes wrong.

There is no practical and inexpensive way to prevent slow oxidation of bituminous coal. Because temperature of the pile is a factor, coal stored in warm weather has a head start in reaching ignition temperature. To avoid fires, store coal so either air circulates freely or is excluded from the pile. The first-in first-out method of inventory control helps reduce storage time but is not always practical, particularly with large volumes of coal. And remember that conditions leading to ignition are promoted by dropping coal from a height, such as discharge from a belt conveyor or elevator.

B B Brown, *Hagerstown, Md*

Cut off air supply instead, via compaction

Diesel oil is not recommended for fighting coal-pile fires. Even if the coal is well compacted and segregation is minimal, oil may flow through air passages and be heated to ignition temperature, aiding coal combustion. A better way to fight these fires is compaction with a bulldozer to cut off air supply. Placing blocks of dry ice in the zone above the fire will also help. Do not dig into the fire area; the fire will spread when the coal bursts into flame when exposed to the air.

In setting up a pile, clear away any accumulation of refuse, vegetation, and organic material. If drainage is poor, locate ditches along the sides of the pile. A base of fine coal spread on the mat will aid recovery of stored coal. Slope and compact tops of successive layers to keep rain, snow, and ice from penetrating the pile. Periodic check of pile temperatures gives ample warning time to prevent a fire. If temperature reaches 150F, remove the coal for immediate use, or segregate it and wet it down.

H B Wayne, *Jamaica, NY*

Temperature monitoring is the key

I have not heard of the method of spraying diesel oil on a smoldering coal fire, but my best guess is that large enough quantities might smother the fire and also present an inherent danger, because diesel-oil flash point can be as low as 126F, which is below coal's volatile release temperature. Application of water

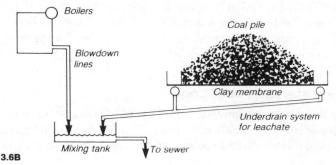

is another course, although water on the pile will mix with sulfur in the coal to form sulfuric acid, a serious pollutant. Collecting the low-pH leachate and mixing with high-pH boiler blowdown will give a sewer-disposable fluid (see sketch, 3.6B).

If a fire is definitely in progress, telltale wisps of vapors will appear, especially on cold mornings, as heat inside builds to around 150F. Don't panic— simply work toward the hot spot in the normal course of fuel consumption, and burn the warm material as soon as possible. If the temperature gets to about 400F, the material should be dug out, spread on the ground, and covered with a fire-resistant canvas. CO_2 fire extinguishers discharged into the covered mass will then eliminate air trapped under the canvas, and the fire will go out. Water lances, the last resort, will work reasonably well, but be sure to control the resultant leachate. K J GARLOCK, *Ithaca, NY*

3.7 An effective cold-climate fuel tank

We are thinking of installing a 15,000-gal Bunker C tank to serve a test-facility station in an environment with a four-month winter averaging 0F temperature. The proposed tank will be flat-bottomed, 11 ft in diameter by 25 ft high. There will be transfer pumps from tank to service heaters in the boiler room, about 30 ft away. Heating-steam at 15 psig is available in quantities of 500 lb/hr from a line 20 ft away. Filler-truck access will be 15 ft from the tank. We estimate that oil consumption from the tank will vary from 6 gph at night to 94 gph in the daytime.

We are considering both aboveground and underground installation, and since the tank could be exposed to a chill factor because of 40-mph winds expected when the temperature is as low as − 40F, we would like to get comments from POWER readers on several design points.

First, should the tank be aboveground or underground? Is a suction heating coil adequate? If not, what type and height of tank heaters would be required? What weatherproof insulation would POWER readers recommend for the tank? Should the filling connection be at the top or the bottom? If it is at the top, is there a practical way to drain the filling line (2-in., about 40 ft long) before it cools off after filling? Finally, what kind of fuel-level meter will be responsive at subzero temperature?—AAN

Recommends compromise on half-buried tank

In the − 40F environment of the proposed installation, a partial underground setting is advised. Earth excavated from the 6-ft-deep hole to form a spandrel-shaped collar will cover about 40% of total tank depth. This will give resistance to wind pressure when tank level is low, and will help reduce heat loss during cold weather. Sketch (3.7A) shows an oiled-sand setting for the tank bottom, although it can be set on a concrete pad if enough asphalt-felt sheet or tough mastic cushions irregularities in tank and concrete surfaces. Sand backfilling is advised to prevent insulation damage, and subsurface drainage is imperative to avoid frost heaving.

Once the tank is set and plumbed, concrete pit and trench work follow. Careful forming, pouring, and vibration of the concrete are necessary to avoid deforming the tank wall. The area of the wall where the pit form joins should be temporarily braced. After forms come off, tank insulation can go ahead. The sprayed-on urethane-foam treatment shown in the "tank setting" sketch will reduce heat loss to around 10% of that for a bare tank. The loss will be about 25,000 Btu/hr when the tank holds 15,000 gal at 140F and the outside temperature is −40F. About 40 lb/hr of steam at 15 psig will be needed—a small percentage of the 500 lb/hr available.

With insulation and backfilling finished, the piping can be started, as shown on the "tank fitting" sketch (3.7A). The internal through-the-top filling line is favored to prevent aeration of oil when the tank is being refilled, but the line

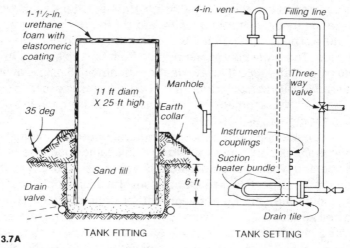

3.7A   TANK FITTING   TANK SETTING

must be braced to prevent excessive movement during filling. A vent, or vacuum-breaker hole, is required to prevent back-siphoning when the filler valve is opened.

The three-way valve allows pumping versatility. If oil must be removed from the tank, the valve can be set to draw from the bottom. After filling, the valve can be opened to draw off to a bucket the leg of fuel in the vertical line. A suction heater of 50,000-Btu/hr capacity will bring cooler oil up to operating temperature when midwinter deliveries are made. And finally, all-iron or -steel valves are resistant to corrosion from No. 6 oil. Do not use brass, brass-trimmed, or even Monel valves in the oil lines.

Second sketch (3.7B) shows a method of installing the rest of the steam, oil, and condensate piping. A sump to collect tank settlings and pipe leakage is shown in the 5-ft × 7-ft concrete pit, 7 ft deep. Arrangements of pumps, filters, and valves will depend on local conditions, but a typical assembly is pictured. With oil heaters on the boiler front, auxiliary heaters may not be necessary. Insulating the line from the tank will keep the temperature in the pipe within a few degrees of the storage-tank temperature, even at the low 6-gph night rate. Tracing the line with 15-psig steam will heat the oil above the tank temperature.

B B Brown, *Hagerstown, Md*

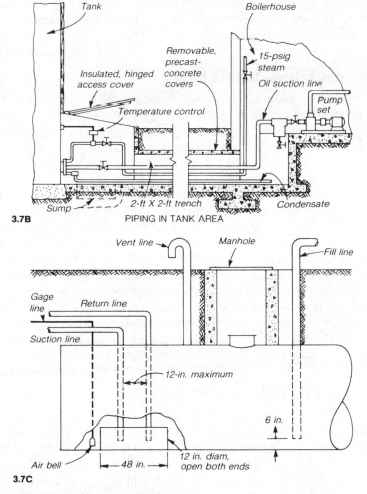

3.7B   PIPING IN TANK AREA

3.7C

A hotwell in the tank will help

An underground tank would be better here, because of the low ambient temperature, the wind-chill factor, and the ease of delivery by gravity or pump. No dike is needed, either. If possible, AAN should install a hotwell in the tank, as 3.7C shows, so the heated oil not being used by the boilers returns to the tank in the vicinity of the suction. From consumption figures, about 5000-6000 gal will come in per week. Temperature is usually about 130-160F at point of delivery. For the pipe coil, 10 ft of 1-in. pipe per 1000 gal of oil is a rule-of-thumb figure that should be adequate for underground tanks.

If an aboveground vertical tank is chosen, the tank should be wrapped with a 1½-in. fiberglass blanket and sheathed with an aluminum jacket. The fill line should be at the bottom, with a check valve. Filling from the bottom serves two purposes: The head pumped against is less, and the line to be blown clean at the end of a fuel drop is much shorter. Finally, a dike would be needed, too, to contain the tank contents in case of accident.

G PETERS, *Thorold, Ont, Canada*

Horizontal, buried tank is best

Install the flat-ended tank horizontally, below ground. Any suction lift over 20 ft will give problems and use excessive energy. Cost of digging will usually be less than insulating tank and piping. Also, keeping the oil warm will be easier. Locating the tank top 1 to 2 ft below the frost line and bringing piping into the building below grade will assure fairly constant temperature in the tank, preventing inside condensation.

Run the vent line parallel to the transfer lines and up the wall. Check codes if building openings are nearby. Make sure the vent top is above anticipated snow level. AAN might be better off to recirculate part of the pumped oil back to the tank after heating, rather than heating the oil in the tank. This will eliminate problems with submerged heaters, and will also save on first cost.

One fuel-level meter that's accurate under most conditions is a graduated stick and sounding tube, in a spare pipe connection at the tank top. This also serves as an access port to remove sludge and water from the tank bottom. If snow covering the opening is a problem, a simple hand-pumped manometer-type level indicator should suffice. To monitor fuel consumption more closely over short periods, a liquid meter in the supply line to the burners is effective.

T A RISO, *Richmond Hill, NY*

Install a hydrostatic gage as meter

Experience shows that installation of outside storage tanks should be discouraged, because this type of storage is sensitive to sudden changes of temperature. The underground tank is better. Heaters should be installed around the tank if law requires the tank to be in a concrete or concrete-block vault. The internal oil heater should be of the suction-bell type.

The ideal fuel-level meter is a hydrostatic gage. These gages are sufficiently accurate for use with up to ¼ mile between tank and indicator. To select the gage, AAN must know the depth or diameter of the tank, the specific gravity of the oil, and the length of the communicating tube.

C BERNARD, *New York, NY*

3.8 Blend fuels to get sulfur content?

A hospital in our state has a boiler plant with three 20,000-lb/hr packaged units, 10 years old, burning gas and No. 6 oil. Gas use has declined, but steam demand has increased, so oil consumption is up, to about 400,000 gal in 1977. Air-pollution regulations limit firing to 2.1% sulfur at 100% capacity or 2.7% sulfur at 70% capacity. Present oil ranges from 2.4% to 2.7% sulfur and is being fired under a short-term easement.

The state environmental authorities have moved to force the firing of No. 2 oil (0.5-1.0% sulfur), but limited availability of No. 2 requires accumulation and storage in summer to meet winter needs. Present storage consists of six very old 40,000-gal bolted steel tanks. They will hold No. 6 oil but leak badly on No. 2.

Our compromise solution is replacement of two of the worst old tanks with new 60,000-gal tanks for No. 2 oil. Blending of the low-sulfur light oil with high-sulfur heavy oil will give oil of medium sulfur content and viscosity. Our problem is: How can we blend the light oil with heavy oil on an automatic basis to give the desired sulfur content within ±0.05% sulfur? Can Power readers tell me if we are asking for too much, and also if there will be any saving over buying oil to spec?—UNS

Sulfur analyzer can control system

Here is a method in which a continuous sulfur analyzer resets the blender system (sketch, 3.8A). The mixer is a static inline type. Low-flow pump returns may be necessary, depending on the sulfur content of the fuel and the amount of fuel burned.

R E RYDER, *Munster, Ind*

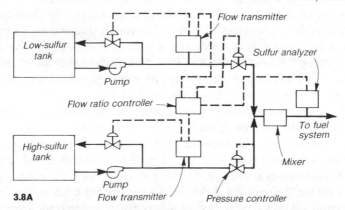

3.8A

Blending at the burners is advised

Chances are that there will be a saving, depending on oil price, cost of new tanks, and some other figures not given in the problem. Blending at the burner is advised. Steam is necessary to atomize the heavy oil. The light oil should go in the burner ahead of the steam, and the steam flow should be adjusted according to appearance of the flame. This method eliminates additional tank and lines for the blended oil.

The worst case is the reduction of 2.7%-S oil to 2.1% by adding 1%-S No. 2 oil. A gallon of 2.7%-S oil plus 0.545 gal of 1%-S oil gives 1.545 gal of 2.1%-S oil, assuming no volume change on mixing, although such a change may occur. If the No. 2 oil is below 1% S, less will be needed. Whether the ±0.05% range in sulfur content is reasonable depends on how reliable the oil-analysis sampling is and how closely the flow rate or pressure of the two oil lines can be controlled.

F G Norris, *Steubenville, Ohio*

±0.05% control is a difficult goal to achieve

There is no cheap and easy way to assure ±0.05% control of sulfur content in the blended fuel. Specific-gravity variations in the usual oil specifications are large enough so that straight volumetric mixing can exceed the desired limit, although that method is cheapest and relatively simple to install, operate, and maintain.

For simple volumetric mixing, the maximum specification allowances must be made for the two fuel streams being blended. At the high end, 2.7% for No. 6 and 1% for No. 2, a mix of 65% of No. 6 to 35% of No. 2 is required for the 2.1% final sulfur content wanted at the 100% firing rate. In design considerations, remember that it is pounds of oil, not gallons, that is the important factor. With the simple volumetric mixing method, a 1.73%-S mixture would result if the fuels had minimum values for sulfur: 2.4% and 0.5%. This would be far below the allowable limit of 2.5% sulfur at 70% firing rate.

Two systems for relatively simple mixing suggest themselves. The first is mixing of oil in a day tank, feeding the burners from here. Two tanks are necessary, one for current use and the other for mixing the next batch. The system will be reasonably accurate if the sulfur contents are known (sketch, 3.8B). A second volumetric method would be to feed the No. 6 and No. 2 streams to a double-pump arrangement with a variable-speed drive, to furnish oil to the burners on a demand basis (sketch, 3.8C).

If neither expense nor complexity is a concern, a more sophisticated system can be devised with instrumented control. The density is measured for each stream, and individual variable-speed pump motors maintain accurate flow metering of each pump (sketch, 3.8D). If feed rate, pressure, and sulfur content of the two streams are supplied to the microprocessor, the accuracy can be controlled within the required ±0.05%. It appears that UNS is a victim of the environmentalism prevalent today, which seems willing to trade some possible respiratory problems for shortages of domestic heating capabilities and illnesses caused by exposure.

B B Brown, *Hagerstown, Md*

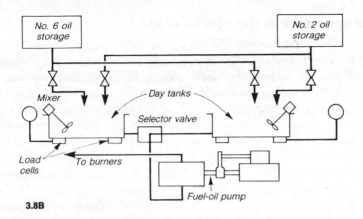

3.8B

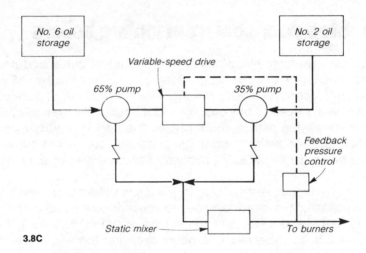

3.8C

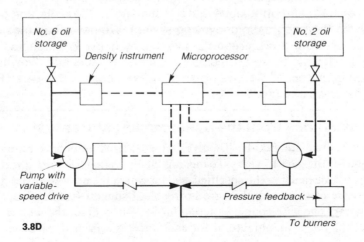

3.8D

Buy to spec rather than try to blend

We had considerable trouble trying to blend heavy and light oils. We finally gave up the effort, and rely now on purchase to spec alone. Lack of stability of the blend was one source of trouble. Separation of blend components gave much difficulty with burner settings, for example. Another problem was the high cost of doing the blending in small lots in tanks. Mixing took a long time, and heating the oils before and after blending was troublesome. We were never able to predict just how long a blending operation would last, either, and we had to add to our tank capacity because of this. UNS would be well-advised to search for a source of fuel oil to meet his spec, even if he has to compromise a little on sulfur requirement.

J D Benson, *Milwaukee, Wis*

3.9 Stop poor coal from fouling boilers

In our plant, we have two 300,000-lb/hr industrial pulverized-coal-fired boilers, which have been doing a great job for us over the past 15 years— that is, until recently. Our regular coal supplier went out of business, and we have been forced to buy coal on a spot basis whenever we can find it. And we are paying astronomical prices. The part that hurts most, however, is that, even with the premium price, this coal is not burning well, and the boiler is not behaving properly. (I'm sure it's not the fault of the boiler.)

We are getting all sorts of slag formation in the furnace, and fouling in the gas passes. This is reducing our steam-production capacity. It seems that the soot blowers are no longer doing a good job of cleaning. Furthermore, there is an unbelievable amount of clinker forming in the bottom-ash and flyash collection hoppers. The material forming in the flyash hoppers is so hard and obnoxious that it is sticking on the walls, and periodically jams the pneumatic conveying system. We never had these problems before. Can Power readers give us any help? What should we be doing to alleviate these problems? Are there any fuel additives that can be used in our case? Can we do anything with our sootblowers? Should procedures be changed?—SR

Get assistance from the boiler manufacturer first

SR says his two 300,000-lb/hr pulverized-coal-fired units have been doing a great job for the past 15 years. To me, this indicates that SR's organization has consistently been using the specified coal for which the units were designed. In the absence of that coal, they are in trouble. I suggest from experience that there is only one wise course of action for SR—to contact the service department of the boiler manufacturer for assistance. A service department having the original design data and a thorough knowledge of combustion behavior can really help. Chapters have been written on this subject, but the manufacturer's service engineer can offer a concrete constructive proposal.

H W Moore, Jr., *Trumansburg, NY*

Re-establish fuel standards

As SR notes, his problem began when he changed coal suppliers. He is obviously being furnished fuel for which his steam generators were not designed. Coals are classified according to their ash fusibilities as follows:

■ Class 1, fusibility range 2600F to 3100F. Such ash is refractory and cannot create problems.

■ Class 2, fusibility range 2200F to 2600F. Has medium fusibility, with minor slagging problems.

■ Class 3, fusibility range 1900F to 2200F. Easily fused, and causes excessive clinkering and slagging unless used in furnaces specifically designed for it.

Consequently, the fusing temperature of coal ash is an important factor in specifying fuel for pulverized-coal firing. If the boiler furnace is designed for high-fusion ash, it will not perform well with low-fusion ash. Difficulties such as rapid deterioration of furnace walls, fused masses of slag in the ash-pit, and large amounts of slag carryover onto the furnace walls and into the gas passages will be the result. Generally, furnaces designed for a poorer grade of fuel will perform well with better grades, but the reverse is seldom true.

To answer SR's questions: The most obvious route is to re-establish his fuel standards to meet those for which his two steam generators were designed. This may be costly in the present sellers' market, but at least one corporation has found it worthwhile. Another way to reduce the problem is to fire with more excess air. This will reduce flame temperature and ash fusing, but will also reduce unit efficiency.

Perhaps SR could balance his present operating cost against increased fuel price without problems. Increased fuel cost, or increased consumption and decreased efficiency with the excess-air method, are the alternatives. SR is a victim of the current fuel problem, like so many others.

B B Brown, *Hagerstown, Md*

Inject steam into forced-draft fan?

The ash-fusion temperature of the coal is evidently lower than it should be. Whether ever done before or not, one possibility for raising the ash-fusion temperature might be spraying the coal with an MgO solution, which raises the value for oil. Water-washing the boiler with hot water at a high pH is a good method for cleaning out slag and achieving a good metal condition. The metal should then be painted with lime whitewash, because the slag doesn't seem to stick to that as readily as to bare metal. Dry the furnace out slowly after this.

Here is a trick that was used on chain-grate-fired units after they were discovered to slag more in winter than in summer: Low-pressure steam was injected into the forced-draft fan discharge to raise the relative humidity. Running a 1½-in., 5-psig steam line into the fan discharge of an 80,000-lb/hr unit increased time between outages from three weeks to three months.

J L AIKMAN, *Montreal, Quebec, Canada*

High excess air reduces slagging

We experience wall slagging because of bad coal, just as SR does. We maintain our sootblower and related equipment carefully, and I assume that SR does too. To reduce waterwall slagging, our operators run at higher excess air than for normal operation. We tilt burners up and down frequently to avoid build-up at one area of the walls. SR should remember to reduce the air back to normal setting when the coal improves again, however. Although SR may not have a superheater, other readers should be cautioned on the loading of the upper burner. Tilt and burner loading control superheater temperature. Excess air, of course, wastes heat to the stack.

E C CUBARRUBIA, *Avon Lake, Ohio*

Wet-bottom operation might help

SR obviously has been getting coal with high ash content and low ash-fusion temperature. He can reduce his problems by several measures, although he cannot eliminate them completely. If he continues use of the present fuel, the life of his boiler, as far as waterwall and superheater tubes are concerned, will be affected.

SR could change his dryash-disposal system to a wet type, with bottom ash molten as in a slag-tap furnace, provided the furnace design is capable of this. Directing a burner flame downward at the pool of molten slag will also help increase travel and combustion time. The capacity of the ash-handling system for both the bottom wet part and the flyash part will have to be increased.

R K JAIN, *N Delhi, India*

Heat release rate is clue

SR's problem is one we will be hearing more of in the next century, as plants try coal again. Similar ones were faced years ago, when efforts were made to fire boilers on fuel for which they were not designed. And that is what I suspect is at the heart of SR's difficulties. Furnace heat-release rate is one good guide to what is happening in SR's steam generators. If he will figure heat-release rate in Btu/ft³/hr, he will have a clue to the direction he must take on his fuel. I am sure he will find a rate too high for his furnace.

In the past, he got by with premium coal, and therefore saved on capital cost. Now, with premium coal hard to find, he must sooner or later face up to a cut in boiler rating. This means more capacity to be installed at present high prices, but there is no way out unless he finds a new source of fuel.

D Y ROBERTS, *Chicago, Ill*

4 Plant Electric Systems

4.1 Tips for synchronous-motor drives

Our plant is planning to install several large reciprocating compressors, each of which will require 1000-hp drives. We have been advised to specify synchronous motors for these pulsating loads. Up to the present, we have had no synchronous drives in the plant and are somewhat hesitant to go ahead with the project because of difficulties that we have heard of in connection with this type of motor.

The total horsepower, 4000 in all, would be a sizable part of our nominal load of about 16,000 kW. One element compounding our hesitation to go ahead immediately with a specification for synchronous motors is the lack of information available on the line layout in general in the plant. We would possibly have to know much more than we do now about short-circuit capacity and about other loads. What has been the experience of Power readers with the introduction of synchronous-motor drives into an existing plant? What kind of engineering should we do before deciding on this type of drive? Is it necessary to consult with our utility at any stage of the intallation? — KRE

No trouble at this plant using compressors

Our powerhouse has a bank of three air compressors and another bank of three

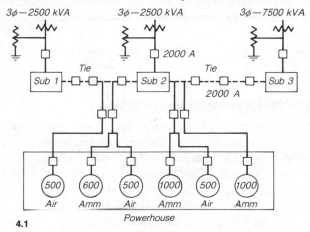

4.1

ammonia compressors, all equipped with synchronous motors varying from 500 hp to 1000 hp (schematic). The two-stage, 180-rpm, 2300-V ammonia units went in during the period from 1926 to 1929. The two-stage, 300-rpm air machines were installed in 1928 to 1950. Total plant electrical load never exceeds 5000 kW, and even with capacitors, our power factor never surpasses 90% for the whole plant.

Synchronization of the individual motors can be slightly adjusted to prevent annoying pulsations, which may rattle loose windows in nearby buildings. Recently, we had to install a strip vinyl curtain wall to reduce the noise level to 85 dBA at one end of an ammonia compressor near the operators' office area. Because of excellent maintenance through the years, the equipment is still in good operating condition. We occasionally cannibalized surplus compressor units for spare parts. The only major maintenance in the past 15 years on the synchronous motors was a rewinding of the dried-out field coils on one motor driving an air compressor. Monthly inspections are made on carbon brushes, and every two years the motor coils are solvent-cleaned.

D G Law, *Shawinigan, Quebec, Canada*

Power-factor improvement seen

The synchronous-motor drive for recip compressors is an old and proven technology, with no special fears or concerns. Part of the good news for KRE is that the drives will improve power factor, giving more distribution-line capacity on low-p-f branches and lowering some of the utility charges. Aside from p-f improvement, we have never experienced significant changes. Control of excitation of the synchronous motors is a must, but the vendors' service manuals and field engineering staffs will advise KRE on that part of the operation. The number of vendors of 1000-hp motor-driven recips is not large, and they are all willing to supply as much engineering help as the buyer wants. Remember that their business is now quite competitive.

A survey of the internal-distribution system of the plant is the responsibility of the plant's engineering group. Line sizes and other data are needed to find capacities. Consulting the local utility may produce helpful suggestions, too, especially with the on-site distribution concerns. The local utility also may have some service-line and equipment modifications because of the higher meter load.

B B Brown, *Hagerstown, Md*

Current pulsation needs user review

Synchronous motors act as generators when a fault condition develops. This will raise the level of available sustained and momentary short-circuit current. Review existing switchgear interrupting duty and associated buses and cabling to see if they can withstand a higher level of fault current. KRE may have to reset protective relays of the affected network to get new trip and time values. A relay-coordination study for selective tripping is in order, too.

The pulsating load will cause both current and voltage pulsation. The motor inertia constant and the driven load also need consideration. The excitation characteristics of the motor itself need to be examined, too. Compressor C factor, described in NEMA MG1-21.88, gives numbers that determine the inertia constants for limits of 66%, 40% or 20% current pulsation. Once the acceptable limit of current pulsation is chosen, the inertia constant of all the rotating parts, such as motor, compressor, and flywheel, becomes calculable. The effect of objectionable current pulsation on equipment connected to the same source must be studied.

A I ORLOFF, *Stanton, Calif*

Look at natural vs forced frequency, too

A synchronous motor draws a pulsating load from the power line when driving a recip compressor. This produces voltage variations or flickering of lights. If the frequency of the voltage change is high, then the voltage change needed to cause noticeable flicker is less. For the start of large synchronous motors, the electrical system must have enough short-circuit capacity to prevent large drops in voltage during inrush. In other words, the supply line must have relatively low reactance, and so the actual voltage variations can be higher before noticeable flicker occurs.

A curve of current pulsation vs C factor is necessary to find out what a given compressor will represent to the power line in terms of current pulsation. The power curve on which the curve is based comes from modern computer programs. The compressor factor C is a function of inertia, speed, frequency, and a synchronizing-torque coefficient. The last-mentioned term depends on voltage and frequency of the power system, load magnitude, operating power factor, power-system impedance, and torque-pulsation frequency. The compressor-motor designer will select an inertia value (WK^2) for the motor, either adding a flywheel to the motor or increasing rotor diameter.

A large WK^2 increases pull-in torque, usually meaning higher inrush currents. Thermal capability of the starting cage winding must allow for the extra kilowatt-seconds to be stored in the winding. Natural frequency is another consideration. KRE should calculate it to make sure the forced frequencies set up by driven-machine impulses do not coincide with the motor's natural frequency.

R J POTTS, *Atlanta, Ga*

Efficiency can be better than for induction motors

Where induction loads predominate, synchronous motors may be employed to advantage, even though relatively low-cost static capacitors are available for power-factor improvement. In addition, when load and speed are nearly constant, synchronous motors are usually more efficient than induction motors. Voltage regulation can be closer if the motor size is 500 hp or more.

In recip equipment, the varying load torque may cause pulsation in armature current and variation in supply-line voltage. NEMA Standard MG1 21.84 limits armature-current variation to 66% of maximum rated load. For system stability, the coefficient of speed regulation C should not exceed 0.003 figured from $C = (N_{max} - N_{min})/N_m$, with N the motor speed and N_m half the sum of maximum and minimum motor speeds.

For best results, the compressor manufacturer should furnish motor, compressor, and flywheel as a complete package, and should have full responsibility for preventing electrical and mechanical problems. Size and location of the flywheel are especially important.

H B Wayne, *Jamaica, NY*

4.2 Motor speed wanders after cold start

We have a large Ward-Leonard drive system for a rotary frequency converter delivering power to a test rig (sketch, 4.2A).In normal operation, the equipment is stopped for several days to a week. Then, on startup, the variable-speed set must accelerate to maximum speed, 514 rpm. At that point, the dc motors are running with fields weakened to about 20% of rated strength. Speed control is by a computer that issues current commands to the generator and motor shunt-field exciters. The speed-control function is very like conventional analog control systems, and includes an adaptive gain function for response in the weak-field region.

Operation is very unstable near 514 rpm when the machines start cold. Even with the controller turned off (shunt fields fixed), speed wanders by several rpm. Usually the oscillation increases in magnitude, accompanied by an increase in dc loop-current oscillation, until the equipment trips off-line from excessive loop current. No tuning adjustment has been

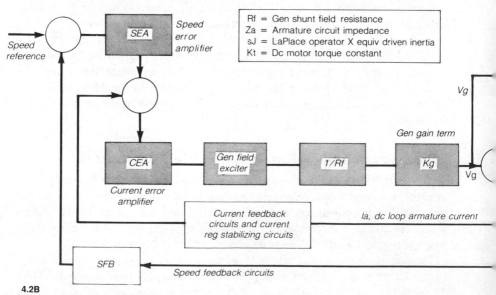

4.2B

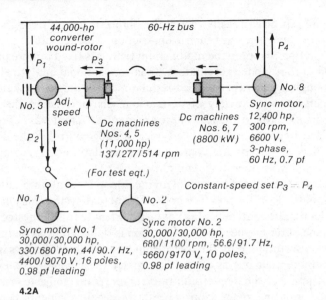

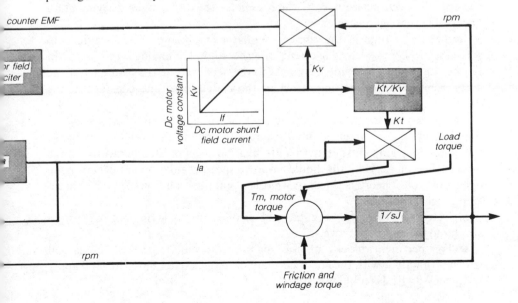

4.2A

able to stop this. When the equipment is warm, it is easy to tune the speed controller. Have POWER readers any engineering explanation for this phenomenon?—AGC

Analysis beats trial-and-error techniques

The clue given about wander of speed with the shunt fields fixed calls for additional investigation. If both motor and generator fields are fixed, it is not surprising to find the speed wandering. In a mode like that, closed-loop regulation is defeated, and speed can easily vary at the slightest change in the characteristic of any component, such as brush drop, friction, windage, etc.

The assumption that only the motor shunt field was fixed is more promising. If operation in that mode is oscillatory, then conventional analytical methods apply. Although a generalized speed-regulator block diagram (sketch, 4.2B) does not include every variety of dc-machine characteristics, the one here should be adequate for an answer. There are many different ways of representing a dc Ward-Leonard control. The typical analog control in the diagram is equally applicable to computer control, where regulator error is calculated instead of summed in an operational amplifier.

The illustrated regulator consists of an inner loop for armature-current control and an outer loop for speed control. The actual system may have more regulating loops, although they may not be pertinent to this suggested solution. Obviously, the outer or speed-regulating loop is defeated if, as deduced from the problem description, only the motor shunt field is fixed. AGC's indication that armature current begins to oscillate under these conditions, eventually leading to tripping of the loop-circuit interrupter, points to basic instability of the current-regulating inner loop. The forward-gain elements contain two transfer functions that are subject to change with temperature:

- Generator shunt-field resistance (R_f).
- Armature-circuit resistance (R_a).

Depending on the dc machine's design parameters, these values of copper resistance may vary from 1.21 per unit to 1.36 per unit, going from room temperature to operating temperature, with a value of 1.0 per unit at 25°C. Result is an increase in circuit gain from cold to hot conditions. Usually, an increase in gain accompanies a decrease in regulator stability. The fact that the opposite occurs here may be the reason for the difficulty in arriving at an acceptable tuning adjustment.

Perhaps a change in the current-regulator stabilizing circuit would help solve the problem. A lead-lag circuit of either digital or analog type must be in the current loop. The evidence points to the need to tune the lead part of the stabilizing circuit, offsetting effect of the closed-loop time constant of the generator-field exciter.

Insufficient separation of lead and lag time constants is another possibility. Any tuneup of the current regulator ought to include an attempt to increase the closed-loop gain and get further attenuation of copper-temperature change effects. The increased gain should improve speed-regulation in normal-mode operation, too. Theory is fine, but what is a practical solution? Here's a three-step general approach:

- Determine the actual regulator block diagram, by laying out the system similar to the diagram above.
- For pertinent blocks of the diagram, formulate the mathematical transfer functions with the aid of dc-machine data, exciter data, etc, and apply frequency-analysis techniques to the results.

■ Test the drive under locked-rotor conditions, with small-step-response techniques. This condition can confirm the transfer functions as required and tune up the current regulator. When gain and response are satisfactory, return the drive to service.

One final caution—on how *not* to troubleshoot drives such as this one. Trial-and-error techniques will probably not succeed and might damage components.

J H SMITH, *Somerset, Ohio*

Warmup period may be only answer

I assume AGC has tried to get help from the original vendor and has failed. In that case, perhaps the only solution is to make sure that the drive control is warmed up enough before each start. This is generally necessary in any setup where heavy currents are involved. The heating effects do not stabilize until all components are at temperature.

T W ARONSON, *Milwaukee, Wis*

4.3 Motor termination, winding failures

Our fertilizer plant's motor inventory includes 21 3.3-kV motors ranging from 200 hp to 2500 hp, and about 600 400-V motors of various sizes. All are the squirrel-cage induction type with TEFC housing, and properly rated aluminum cables feed all of them. In the past eight years, we have had several aluminum-cable termination failures and motor burnouts on the 400-V units. We have scrupulously adhered to the usual precautions for terminating aluminum cable, such as inhibiting paste and bimetallic washers. In addition, we carefully seal the terminal boxes to keep out moisture, because of the dusty, corrosive atmosphere.

Also, motors that had been running continuously for several months suddenly developed winding failures. One 150-hp motor failed after a couple of hours' run, even though it had shown good insulation resistance before starting, and had run satisfactorily for several months prior to the interim stop. What has been POWER readers' experience with recent failures of these types? Is there a connection between termination failures and motor burnouts? Are there effective test methods to warn of impending trouble from termination and winding defects?—ML

Torque curve and starts are key

ML could check the NEMA standards on motors. Some of these recommend a higher horsepower rating in applications requiring overload capacity. This will protect against excessive temperature rise and give enough torque capacity. Also, ML should check margins between motor speed/torque curve and load speed/torque curve. Motor life is affected by the number of starts, and any recent change of motor controls, such as to solid state, may contribute to the problem.

W S KOZINSKI, *Des Plaines, Ill*

Look for intermittent ground fault

ML's plant fertilizer dust could be one of several possible answers—corrosive and electrically conductive salt. Another and very probable cause is an intermittent ground fault. Such a fault, combined with the electrical-system capacitance, can set up an impressed dc offset voltage spike with magnitude of six to eight times phase voltage, or approximately 3.5 kV on 500-V insulation. This could readily puncture 600-V insulation. In addition, the spike can easily travel over and through various distribution systems.

ML's reply to this could be "Wait, I do have a ground fault, but it has never tripped the main breaker." This could easily be true, because what is causing the problem is a voltage spike, not a current spike, so the ground-fault system doesn't see it. ML's solution will probably be some form of a high-resistance ground system, such as a zig-zag transformer with a grounding resistor. With this installed, ML can begin a hunt for his intermittent ground fault with signal generators, or other methods of ground-fault detection.

J R SULLIVAN, *Winchester, Va*

Corrosion from fertilizer is a danger

Arcing at the terminals can definitely cause a winding failure similar to a low-voltage failure. Possibly ML's arcing is worst at peak loads and maximum vibration, which would further aggravate the situation. Transients and high-frequency waves generated at the arcing terminal could cause puncture of the winding insulation, because of higher-than-normal voltage per turn, or could

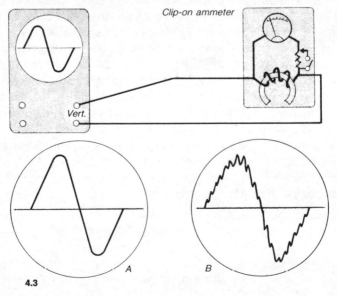

Clip-on ammeter

Vert.

A   B

4.3

magnetically destroy the insulation by magnetic pulses. Around fertilizers, almost any form of corrosion could occur on the aluminum/copper junctions in terminal boxes. Urea and phosphoric acid are causes. Neutralizing substances could leave undesired water, too. Desiccants—a possible cure—would be too hard to monitor in a closed junction box.

ML should replace the aluminum wiring, if possible, because copper-to-copper withstands more contamination. As a side note, we have found that regular yellow brass stands up to high-sulfur cutting oils, which would quickly eat away pure copper. Brass, rather than copper, in troublesome power and sensing areas can withstand sulfates in some fertilizers. If ML elects to stay with the aluminum wiring cables, he should open all terminations, file and wirebrush them clean, and tighten them after applying nonoxidizing paste. Periodic arc-detecting checks at each motor are then done during operation and without disassembly.

A clip-on probe-type ammeter on each wire, with a sensitive oscilloscope connected to the terminals of the ammeter coil or in many cases to the meter terminals, will show the sine waves in the sketch. A pure sine wave (A) indicates no arcing, but if a fuzzy pattern (B) appears, then either terminal arcing or winding failure is in progress and will cause a shutdown. This test will isolate troublesome areas, give data on spare-motor requirements, and perhaps point out process batches that are especially corrosive to the equipment.

D THELIN, *Chandlers Valley, Penn*

Winding-failure sources are many

ML has apparently been following standard practices for connecting aluminum cable to the copper motor leads on his 50-Hz 400-V units. He should make sure that the connectors are marked "Cu-Al," too. His application of inhibiting paste is proper, but he could also paint the ends of the aluminum conductors with red lead before joint makeup. If motor leads are long enough—which is not usually the case—compression connectors could serve.

There still doesn't appear to be an aluminum/copper connector that is absolutely troublefree, but ML's experience of "several" termination failures on 600 motors in eight years isn't too bad. The many causes of field-winding failures include:
- Low power factor
- Lubricant contamination of the windings
- Low voltage
- Overfusing, with subsequent motor overloading
- Insulation failure
- Repeated jogging of the motor

It is possible for the conditions leading to termination failure to cause a motor-winding failure, too. If a motor ran for several months, then was shut down and restarted, conceivably there was arcing at the terminals because of a loosened connector. A high resistance thus formed could overload the other two phases and cause a burnout. This assumes that the overload protection doesn't function. Frequent starts—10 to 20 in a short time—will cause winding failure even though the overload devices will not sense anything. The windings will become too hot to last. Imbedded thermal cutouts will prevent this mode of failure.

A preventive-maintenance program requiring scheduled disassembly of the aluminum/copper terminations, inspection, cleaning, and reassembly with the best available "Cu-Al" connectors is suggested. Because ML is already checking with an insulation tester—an accepted and reliable means—there is little more that he can do on winding defects.

B B Brown, *Hagerstown, Md*

Check vibration at terminals

Although all motors and driven equipment experience some vibration, smaller assemblies are more responsive to external forces. Aluminum and copper are prone to crystallization from vibration action, and the cables may undergo this at the motor terminals. Locating connections as far away from the motor as available space will allow can limit amplitude of flexure. Winding and insulation failures, too, may stem from excessive vibration. In addition, as motors age, insulation dries out and flexibility decreases. A vibration-measurement maintenance program might be a good idea for ML because of the large number of motors he has.   H B Wayne, *Jamaica, NY*

4.4 Best ways to utilize breaker capacity

Our textile-finishing plant has an electric distribution problem on which we need help. We are adding 50,000 ft^2 to the present 80,000-ft^2 area of the plant. We have a 3000-amp distribution panel with six main breakers. Our system is 480/277 V, four-wire. The six main breakers are: an 800-amp with 772-amp connected and 340-amp average working load; a 400-amp with 392 connected and 175 average; and four 600-amp, with 514, 348, 478, and 282 connected, and 350, 280, 250, and 180 average, respectively. The connected load is the full-load current of all motors, and the average load was read during operation.

Power-company charts show a 1200-amp maximum. Connected load is about 2700 amp, and we will add about 1000 amp. Some of the machinery is shut down half of the time, but we never know where the load will be at any given time, or when we will need to run all machines at once. The manufacturer tells us that a 600-amp breaker is the largest that we can install in our switchgear now. Can POWER readers tell us how best to use the complete 3000-amp service without overloading any breakers and causing them to trip when we want to run 100% of the plant at any time? If a study is needed, what is the best way to do it?—SN

Take advantage of NEC's Exception 4

Most practical way would be to leave the existing service-entrance equipment with the maximum number of disconnects (six) as it is, and install a new service entrance to the plant addition under Exception No. 4 of Article 230-2 of the National Electrical Code. Modification of existing distribution equipment is usually expensive, and the downtime in the existing plant could be inconvenient, too. Changing the utility transformer to one of larger size would probably cause additional downtime for the plant, and also increase the available fault currents to where the interrupting capacity of the existing circuit breakers is exceeded. The utility may be willing to give primary metering to the old and new transformers; this could result in a saving on power bills.

W K PENHALLEGON, JR, *Clearwater, Fla*

SN needs a new service for new load

First order of business in SN's problem is the National Electrical Code, which should be adhered to in design and installation of systems. Stated simply, two basic rules are: (1) Capacity is the connected load as determined by nameplate data and computations set down in the Code book, not by running loads or averages; and (2) capacity of switches, wire cables, fuses, etc, for circuits that have surges (such as motor-starting loads), must be larger than the connected load. These two rules will assure 100% operation at all times, without tripouts.

Adding up SN's connected load gives 2786 amps, so that the remaining capacity of the existing system is 214 amp (panel capacity of 3000 minus connected load of 2786). A new service is therefore required if another 1000-amp load is to be added. Major consideration: Should the remaining 214 amp be used, and the difference in required capacity be provided by the new service, or should the remaining 214 amp be held in reserve and all the needed capacity be provided by the new service?

Transformer for the new service will need a vault if located inside the building. SN should make a study complete with accurate data and calculations, sketches, and drawings. Consultation with an electrical engineer or the power company can provide a system profile and its cost. Starting point is determination of load of all consuming devices. Voltage, amperage, and watts will determine the branch circuit, etc. Then determine the distribution panel, feeder cables, transformers, and switchgear.

R DUQUETTE, *Watertown, Mass*

Try combining loads on breakers

SN could combine loads on two 600-amp breakers. With connected loads of 348 and 282 amps, feeders should be the same size and rated for 600 amps. The excess 30-amp connected load should not create a problem, but if it does, it could be shifted to another panel. This combining of loads would result in two 600-amp spares, and the new plant area could be fed on two separate panels, with separate feeders. Or, after loads were combined as above, SN could install a new 1000-amp breaker next to the existing panel, and either connect to the existing 3000-amp distribution panel as needed or subfeed to a new 1000-amp breaker in the new area. Another possibility: After combining the loads as mentioned above, subfeed from the existing distribution panel to two subfeed panels with main breakers in the new area.

R E JULIAN, *Saratoga Springs, NY*

New panel will give added capacity

A new 1600-amp main beside the old panel (drawing), complete with ground-fault protection, should solve the problem. The new breaker setup can feed a 1600-amp distribution panel. Moving the 400-amp main breaker to the new panel would mean that it is no longer a main. A 600-amp breaker in the new panel would help with some of the added load. Result would be only six main breakers (as the Code says), and plenty of future distribution capability.

L E DENNIS, *Sanford, NC*

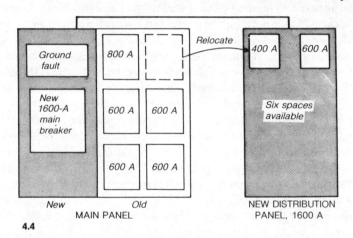

4.4

Don't serve new load from old setup

Simplest and probably least expensive way for SN to provide electric service for 1000-amp additional connected load in his new building would be to establish a new point of delivery near the load center of the addition. This is permitted by Section 230-2 Exception 4 or 5, of National Electrical Code. A panelboard with three main breakers for the proposed load and spaces for three future breakers for other expansion could be installed at the new service. Several expensive problems would be encountered in any attempt to serve the new load from the existing setup:

■ Large conductors would be needed to prevent excessive voltage drop in the 480/277-V feeders between present main panel and proposed electrical loads.

■ A new panel with additional breakers would have to be added to provide for the new load. A connection could be made to the present main bus, but the existing main breakers do not have enough reserve capacity to allow use: Refer to NEC Sections 220-4 (a), 430-24 and 430-25 for capacity requirements.

■ The 3000-amp main bus would have to be increased in size unless the local inspector were to grant a diversity variance; refer to NEC Section 430-26.

■ A single main breaker would have to be installed on the source side of the six existing main breakers if more breakers were added, because no more than six disconnects are permitted at a point of delivery; see NEC Section 230-70 (g).

■ Finally, the utility might have to increase the size of the transformer bank serving the plant. This, in turn, would increase the available fault current and might require replacement of existing main breakers. See NEC Section 230-98.

E F MENIUS, JR, *Hartsville, SC*

Rearrange motors to match loads better

It would appear that all the connected loads are over-motored; that is, the nameplate rating of the motors is far in excess of the actual power required. This is a common situation, and SN should first check the actual load on each machine in operation, using a clamp-on ammeter. The sum of actual loads will give the maximum amperage on each breaker. He will probably find unused capacity on each circuit. Knowing the actual power required by each driven load will allow a planned rearrangement of motors to match the loads better. This will increase efficiency and improve power factor.

M M FROMM, *Cleveland, Ohio*

4.5 Keeping flicker within limits

The question of flicker has become controversial between our steel smelting plant (Philippines) and the local utility. We have a total of 538% impedance up to the arc furnace, with short-circuit capacity of 4.25 MVA. Short-circuit current at the arc furnace is 12,300 A at 200 V. The expected flicker is 4.77%. Our local utility is small: Capacity is 25 MVA at 13.8 kV, with impedance of 10.4% and symmetrical short-circuit capacity of 132 MVA or 665A.

The utility, using a recognized flicker curve, comes up with 0.7% flicker for arc furnaces, welders, etc. Studies made many years ago were the basis for the data. My research, however, shows a 5% flicker limit value set in a AIEE paper study (CP55-802), a 6% voltage dip value in a manufacturer's distribution application manual, and an 8-V average dip limit in IEEE paper 141-1964. I am curious as to what is the percent flicker acceptable to the average utility. Can POWER readers tell me what their experiences have been? At what percent does flicker become noticeably irritating, and is there anything we can do to improve conditions?—GA

Stability of electric utility may be decisive

Usually electric arc furnaces have load and power-factor values as good or better than many other industrial loads. However, supplying power to them can present problems. During meltdown, scrap may bridge the electrodes, approximating a short circuit on the secondary side of the furnace transformer. This causes rapid current fluctuations at low power factor, single phase. When the refining process begins, the steel is in a molten pool allowing uniform arc lengths—available with automatic positioning of the electrodes. Thus stable arcs are obtained producing steady three-phase load at high power factor.

Since total reactance includes that of the primary circuit, distance between power source and furnace influences the voltage dip (sketch, 4.5A). Installing

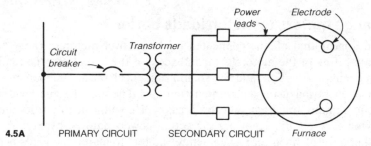

4.5A   PRIMARY CIRCUIT   SECONDARY CIRCUIT   Furnace

capacitors in the primary circuit minimizes the problem while improving power factor. Evaluation of light-flicker magnitude depends on the method of analysis. In the past *percent flicker* was used; today *flicker index* is recommended since it offers a realistic accounting of the wave form of the light output (sketch, 4.5B). Regarding flicker values acceptable to utilities, there are

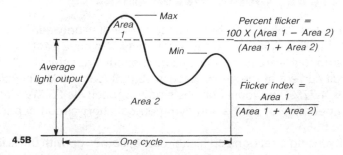

$$\text{Percent flicker} = \frac{100 \times (\text{Area } 1 - \text{Area } 2)}{(\text{Area } 1 + \text{Area } 2)}$$

$$\text{Flicker index} = \frac{\text{Area } 1}{(\text{Area } 1 + \text{Area } 2)}$$

no mandatory criteria. Frequently limits are dictated by the utility system stability, which decreases if the system load increases faster than generating capacity increases—a prevalent situation today.

Unless the Philippines' utility is able to maintain power supply stability, it is very likely GA will have to improve power factor and accept voltage-dip limits specified by the utility. If this is not feasible, consider on-site generation to augment or replace the utility supply. This is particularly true if plant expansion is planned.

H B WAYNE, *Jamaica, NY*

Modify method of arc-furnace operation

The number of starts per day is important to the electric utility. When they occur infrequently, dips limited to 6% or 8% are generally tolerable even though they cause some lamp flicker. (Refer to the Electrical Engineers' Handbook.) Solid-state furnace controls in use today (based on components such as SCRs) do not change power factor. With these controls, the stabilized kW should not vary much until the furnace is up to desired performance. If the furnace could be kept at lower heat during off-load periods, the operating current could be reached at demand periods without much circuit adjustment. The use of series reactance could help overcome the negative resistance characteristic of the arc.

W S KOZINSKI, *Des Plaines, Ill*

Encourage utility's cooperation, work together

Much of GA's problem comes from the fact that his plant load is large relative to the utility system. It would be mutually beneficial, therefore, if the power company system and the steel plant could be viewed as one problem. Very likely GA will have to make some modifications in his electric system, and possibly in his method of operation. But the power company should investigate the actual limits of flicker which its other customers can tolerate.

Studies have been made to determine the amount of lamp flicker that can be tolerated without objectionable psychological effects. Results of such a study are shown in the accompanying curves (4.5C), which have been well known for

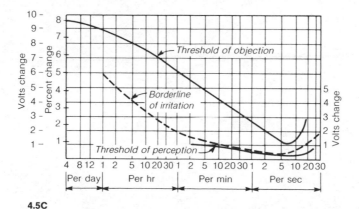

4.5C

years. Because psychological response is an individual matter, wider limits may be used under certain conditions without complaints from the people oc cupying the affected area. The curves are based on incandescent lighting;

fluorescent lamps are less subject to flicker over a range of voltage that is not low enough to put them out. The important thing to observe from the curves is that frequency of the flicker contributes to irritation as much as the percent change (change is based on 120 V). Any system modification that will reduce the frequency will be helpful.

Since GA's problem originates in his arc furnaces, he should consult an expert on this type of equipment. At the same time he should study his electric system in general to discover conditions which may be aggravating the voltage fluctuations originating in the arc furnaces. Capacitors adjacent to the main service can be helpful in smoothing out voltage dips and at the same time improve power factor. They should not be installed haphazardly, however, since they can cause overvoltage or resonance problems unless properly coordinated with other circuit components. With the cooperation of the utility, GA may find it more economical to "cure" the flicker problem at the other end—by paying for voltage stabilizers on lighting circuits in the buildings and residences of affected neighbors.

EMIL WAGNER, *Minneapolis, Minn*

5 Pollution Control

5.1 What happened to RO treatment?

Our station's water-treatment system consists of a reverse-osmosis (RO) setup followed by a mixed-bed ion-exchange (IX) vessel, designed for 1000 gal/day flow rate. Our makeup going to the RO module has about 100 mg/liter of total dissolved solids (TDS), and initially the resulting permeate had about 50 ppm TDS and conductivity of about 70 μmho/cm. The IX section reduced solids to <1 ppm and conductivity to about 1 μmho/cm. These data held for a year. Recently, however, we have been having to regenerate the mixed bed every third day on average, in contrast to the monthly intervals that previously sufficed to maintain low TDS and conductivity.

The IX section looks fairly good and its resins have not deteriorated. This leads us to suspect the RO setup. Another piece of evidence is that the pressure drop across the end RO stage has risen from about 20 psig to 30 psig. What has been the experience of Power readers with similar problems? Is our experience characteristic of all RO plus mixed-bed systems, or should we search for specific causes? Is change in makeup water more likely to cause trouble than deterioration of RO components?—JSW

A graph plotting years of data helps in decisions

One of our plants, in west Texas, has a similar RO and ion-exchange (IX) system. Lake-water makeup to the RO has had up to 2800 μmho/cm (TDS of 1790). Performance loss from this 30-gpm RO also caused more frequent IX regenerations. We are able to attribute the problem of frequent IX regeneration back to the RO unit performance after we made a scatter plot of IX performance. The graph of several years of data let us evaluate number of outflow gallons that the IX produced before regeneration, as compared to the quality of water feed to the IX system from the RO. IX runs that graphed below the curve were reviewed in light of poor IX performance. Any trend that continued to fall below the curve indicated possible IX resin and regeneration problems.

JSW should establish a daily or weekly monitoring program for the RO-IX system. We found that graphing several key parameters by hand or personal computer gave a good picture. Four items are recommended for plotting: pressure of feed to the membranes after pressure controller, conductivity or TDS of RO makeup, conductivity or TDS of the permeate produced by the RO, and percent sodium chloride rejection. Also, get and hold data on temperature, product and reject flow, brine pressure, and pH.

From a good graph (below), observations are possible. Pump discharge

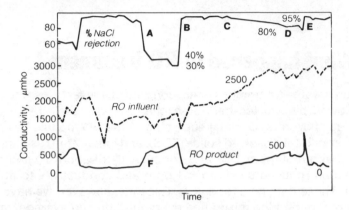

pressure, with filters clean, will vary directly with the product flow and conductivity of the makeup to the RO. A direct relationship exists between RO-makeup conductivity and permeate conductivity. Percent sodium chloride or chloride rejection rate should stay constant, at 87% to 98%. Five years of graphing caught several classical events. At point A there was overnight failure of our membranes, with subsequent rise in product or permeate conductivity at F. New membranes, point B, restored the unit to baseline performance. From C to D was typical slow deterioration of performance, as seen from percent sodium chloride rejection, even though influent conductivity increases from 1500 to 2700. New membranes, point E, again restored the system to baseline.

It is evident that RO systems can have both catastrophic or overnight failure and slow deterioration. We found that catastrophic failure can come from loss of control on the RO pretreatment system where high pH levels (over 6.5) cause hydrolysis of the membranes. A high-pH trip would protect against this type of failure. Although we never could confirm it by chemical tests, we noted correlation between new water entering the lake from rain runoff and changes in RO performance. We suspected that pesticides or other organic contaminants had reacted with the RO membranes and made them fail. We considered activated-charcoal filters upstream of the RO to prevent this, but after evaluating cost and failure frequency, we decided merely to shut down the system for 48 hours after large amounts of new water entered the lake.

JSW's pressure increase may be from fouling or merely from higher TDS concentration of makeup to the RO. Our rule of thumb for cellulose acetate membranes was 10-psig pressure change for each change of 1000 ppm of inlet-water quality. Ten-micron filters are recommended ahead of the RO vessels. Scale formation on the membrane can also cause fouling and subsequent pressure changes. Some types of RO vessels are susceptible to what is known as compaction, which makes the system pressure slowly rise. Although cleaning and backflushing will help, a three to five year life is all that can be expected of most membranes.

J A SCHMULEN, *Ft Worth, Tex*

Apparently the mixed bed is fouled

I think that JSW's mixed bed is rather fouled because the run lengths dropped considerably and the conductivity of "about 1 μmho" is not particularly low. Resin fouling can be caused by foulants from the regenerants or air oxidation of the cation resin. The mixed-bed resins may have to be replaced. Here is a check list for troubleshooting RO/mixed-bed systems:

■ Check influent-water analysis. Usually RO product quality is 10% of the influent. Should RO effluent quality be greater than 10%, look to reject flow and operating pressures.

■ Check acid feed, if any. Acid lowers pH, which is healthy for membrane life. Too much acid, however, will generate extra free CO_2, which RO will not remove. Result is a loading of the mixed-bed anion resin, shortening mixed-bed runs. Consider an added decarbonator to remove this CO_2.

■ Check free CO_2 content of mixed-bed influent. Conductivity measurements, replacing analytical cation/anion tests, are not accurate for estimating mixed-bed loading.

■ Check operating pressures of RO. If permeate is low and operating pressure has been increased to maximum allowable, the RO modules may be fouled. This assumes that waste or reject flow is adequate.

■ Check waste or reject flow. Adjust reject flow to at least 50% of permeate flow; 100% is preferable.

■ Check influent to RO or mixed bed for filterability or silting index to see if foulants are present. A suggested test: Filter a liter at 25°C through 0.45-micron, 47-mm-diameter membrane disc at 14.5 psig or vacuum. If flow is under 300 ml/min for the first minute of filtering, the water contains foulants to RO or mixed bed. If the membrane disc after filtering 1 liter has more than an off-white stain, then the water contains foulants to both RO and mixed bed. Generally RO effluents have flow index over 400 ml/min and are stainfree. Another simple test for foulants in water is evaporation of 1 liter acidified by a drop of 1:3 H_2SO_4 to dryness at 200°C. Then examine the stained salts at glass-beaker bottom.

Consider RO cleansing treatment. If the tests indicate fouling, especially from iron, then a citric acid treatment may be considered. Check mixed-bed resins properly. Ordinary lab capacity tests and mixed-bed anion-resin appearance do not necessarily reveal fouling. Kinetic flow tests or mixed-bed operating-record examination is preferable. New or unfouled resins should rinse down within 5 to 10 minutes, assuming that the final resin mixing operation was correct.

G J CRITS, *King of Prussia, Penn*

Conductivity record is help

If the problem is with the RO portion of the plant, conductivity records should give confirmation. Record daily, as minimum, the flow, pressure, and conductivity of each of the three streams, as well as feedwater temperature, for any operating RO unit. Change in dissolved-solids content of the permeate will reflect in the conductivity. In many cases, mixed-bed demineralizers operating on RO permeate perform poorly because of poor resin separation. The resin tends to clump because of lack of foulants in RO permeate. For good resin separation, soak for an hour in 48°C saturated sodium chloride.

Remember, however, that the anion resin will float. Then rinse the salt solution from the resin in upflow direction and double regenerate without remixing at the end of the first regeneration. To provide an artificial foulant, add one cup of clean bentonite for each 10 ft³ of mixed resin. Then air mix and rinse. Increase in differential pressure in the last stage RO section indicates that fouling, probably because of precipitation of sparingly soluble salts, has occurred.

C E HICKMAN, *Cave Creek, Ariz*

Without cleaning, fouling is key possibility

JSW will have to investigate several possibilities. Reduction in TDS level by 50% from 100 ppm to 50 ppm in the permeate indicates that a relatively large brine purge is applied. The high brine purge must avoid quick fouling and concentration polarization effects at the membrane surface. If the installation is run for a year without periodic cleaning of the RO modules, fouling is fairly certain to occur.

This means increased pressure drop, as observed, and more chance of concentration polarization in the dirt layer on the membrane surface, leading to decreased permeate quality and shorter runs of the mixed bed. Hollow-fiber modules are more sensitive to fouling than tubular or flatplate modules. Periodic cleaning to remove silt, organic fouling, iron oxides, or even calcium carbonate or calcium sulfate precipitates at regular intervals determined by permeate quality is necessary. Detergent solutions and/or 2-3% citric acid solution are required, with regular flushing of the modules.

If fouling becomes too heavy, membranes may be damaged and have to be replaced. Once we operated a 260-gal/hr hollow-fiber pilot unit on potable water, and the unit required cleaning at 40,000-gal feedwater intervals. Depending on feedwater composition, the mixed-bed unit may also be a source of the problem with shortness of runs. If the feedwater has high silica concentration, it may slip through the fouled or damaged RO modules. High silica concentration entering a mixed bed that is regenerated only once a month may lead to aging of the silica on the strong-base anion resin. The precipitated, non-ionic silica is practically unremoved during regeneration. This reduces the exchange capacity of the strong-base resin and shortens the run lengths.

A J V PELT, *Hellevoetsluis, Netherlands*

Bacterial growth harms membranes

Check operation and backwash of the pretreatment system. Some literature references suggest that cycle length is improved if the silt density index is kept below 2. Bacterial growth could also damage membranes and severely reduce cycle length. Biocides in the pretreatment could be considered as a remedy. Proper interlock operation is important, too. Some plants bypass feedwater to RO module if pH is above 6.5 and temperature over 95F.

G C SHAH, *Houston, Tex*

5.2 Stopping damage from metal pickup

We have had serious waterside corrosion on both high-pressure and medium-pressure boilers. The corrosion is attributed to pickup of copper and iron from our paper-machine condensate system. For years we have followed the normally accepted standard of water treatment for these boilers. The three afflicted boilers get high-quality demineralized makeup.

We have been adding neutralizing amine, and also filming amines, but the latter have been unsuccessful in preventing pickup of copper and iron. Condensate polishing (anion resins) was ruled out because of the high temperature of the condensate return. We tried a cation polisher (sodium form), but results were disappointing. We would like to hear from paper and pulp mills, and other industries with similar condensate problems. What has been the experience with metal returned in condensate? What is the best way to handle the problem—at the source or in the boiler? And what are the costs and controls of the solutions?—RB

Go with catalyzed hydrazine plus amines

RB's problems are typical of virtually all condensate-return systems. Corrosion caused by copper and iron deposition from system condensate can follow low pH, dissolved oxygen, or erosion in the condensate system. Successful control has been achieved through proper inhibitor feed in the boiler, backed up by the right sodium-zeolite cationic resin. Ideal internal boiler conditioning would be a coordinated program, with catalyzed hydrazine and a blend of neutralizing and filming amines. Filming amines give a synergistic effect, because the catalyzed-hydrazine-promoted magnetite coating of metal surfaces helps bond those amines.

Catalyzed hydrazine converts ferric into ferrous iron, and cupric to cuprous copper, in the boiler. Ferrous iron oxide is more easily dispersed than ferric, so iron returning to the boiler is more soluble and easier for blowdown to remove. The catalyzed hydrazine scavenges oxygen that could cause copper-oxide corrosion. The cuprous form of oxide resists ammonia attack, another advantage of conversion. Boiler-sludge conditioner and condensate polisher are advisable, too. Feedpoint locations for the sludge conditioner and the hydrazine must be suitable. Feed the former directly into the preboiler system or the boiler, and the latter to both the preboiler system and any points where oxygen in-leakage can occur. Finally, start a resin-cleaner maintenance program.

R E Morrison, *Quinton, Va*

Choice of three: chemicals, polishers, filters

Copper and iron pickup is a frequent problem in the condensate system of pulp and paper mills. Deposition of the metals in the boiler can cause tube failures from overheating and corrosion. In addition, copper will start pitting corrosion. Three methods for preboiler control of these two contaminants, in order of decreasing cost and increasing simplicity of control:

■ Chemical treatment requires feeding neutralizing amines to hold condensate pH in the 8.8-9.2 range (tighter than the normal 8.2-9.2 standard). Selected amines are necessary, with a range of distribution ratios or satellite feed stations at critical points. Hydrazine improves corrosion protection by passivating the condensate metal. An iron dispersant added to feedwater keeps iron suspended in the boiler. During startups, dump the condensate until contaminant levels are acceptable.

■ Cation polishers will remove most soluble forms of copper and iron. Efficiency of solids removal by filtration depends on the regeneration process. Resin regeneration is necessary on either pressure drop or contaminant breakthrough. For regeneration, adjust the backwash rate to break up the resin bed and remove solid contaminants. Mix a strong reducing agent (usually sodium hydrosulfite) and feed it with the regeneration solution. The reducing agent will make iron more soluble and enhance its removal from the bed.

■ Electromagnetic filtration is 60-90% efficient for iron removal. Inclusion and absorption in the matrix of solid particles removes other impurities, including copper. The filters are regenerated on iron breakthrough or pressure drop. This filtration technique is growing for condensate polishing in the pulp-and-paper industry.

RB should treat by two or more methods if iron and copper levels are high. Treatment-program cost relates directly to the volume and quality of condensate.

G A Konkel, *Kent, Wash*

Segregate worst streams to save costs

RB's problem is a condensate/feedwater corrosion problem, probably compounded by organic contamination from the paper machines. Hold the pH of the condensate/feedwater cycle in the medium-pressure boiler system in the

8.5-9.0 range by a neutralizing amine. Rigorously exclude oxygen by treating with catalyzed sodium sulfite, injected well upstream of the boiler drum, preferably in the deaerating-heater storage compartment. Stop the filming-amine treatment, which requires care because feedrate is hard to control.

RB is undoubtedly aware that both excessive corrosion products and organic material must be excluded from the boiler. Total iron and copper in the feed-water should not exceed 10 ppb. If the copper can't be dropped below 5.0 ppb, then consider filtering suspended corrosion products. Examine the condensate-return system to see if some streams are more contaminated than others. Segregate and filter only the worst streams. A pressure sand filter, perhaps a mixed-bed type, is advisable. Lab filter tests and equipment-supplier consulta-tions will help. If organic matter is a problem, an activated-carbon filter is in-dicated.

The best place for RB to solve the problem is not in the boiler, but at the source, where cost is minimal. A boiler examination in this instance should focus on:

■ Evidence of pitting or general corrosion in drums, headers, and tubes.

■ Amount and distribution of sludge in downcomer tubes, especially below the feedwater pipe.

■ Amount, location, and character of deposits in generating tubes and wa-terwall tubes.

■ Plugging in feedwater and blowdown lines.

■ Leaks in baffling in steam drums, and cleanliness of steam-purification equipment.

F SCHOEPFLIN, *San Francisco, Calif*

Magnetism is attractive for removing iron oxides

RB might look into an electromagnetic filter to remove magnetic iron oxides from the condensate system. Under some conditions, significant amounts of copper will be taken out, too—as much as 89%—although the conditions for this are unknown presently. Iron-removal efficiencies of 95% are common, however. Cheapest and best location for an electromagnetic-filter setup seems to be downstream of the condensate-collection tank or manifold, and upstream of the condensate polisher. For this location, the captured impurities won't foul the polisher resin or deposit in the boiler and other downstream equipment.

R R CSUHTA, *Barberton, Ohio*

Treat condensate corrosion at the source is the rule

Without more data, it's difficult to reply to RB, but here are some pointers for his waterside-corrosion troubles:

■ Return-line corrosion, by soluble iron or copper, can be treated by pH adjustment or by lessening the ammonia.

■ Return-line erosion, with particulate iron or copper as the agency, sug-gests 5-micron cartridge filtration.

■ Increased corrosion potential from oxygen in the feedwater yields to deaeration and chemical treatment.

■ Heightened corrosion potential from boiler-water dissolved solids calls for surface blowdown.

■ Corrosion at particular sites may be from solids in boiler water. Chemical treatment and bottom blowdown solve this.

Treating condensate corrosion at the source, eliminating metal wastage in the condensate and reducing boiler corrosion, treatment, and blowdown loss, may be the most cost-effective solution for RB's case.

G S MAHALY, *Linden, NJ*

Electromagnetic filters are on-line

If RB should decide to consider electromagnetic filters for removal of iron and copper from his condensate, here are some suggestions coming from recent field trials on a 100%-condensate-fed paper-mill boiler:

■ Because iron contamination is often in the ppb range, guard against sample contamination, or results will be in doubt. Wait at least a half hour between parameter change and sampling. The phenanthroline method has been successful for total-iron determination.

■ Fluid velocity and flow rate are important. About 200 gpm/sq ft is the limit for the filter.

■ In paper mills, 87-89% average total iron removal has been reached, with 50-57% copper removal.

■ Look into hydrazine as a condensate-oxygen scavenger. It can raise the filter's total iron removal to the 96-99% range but introduces environmental hazards.

■ Plan for timely valve shutoff to prevent trapped-solids release from the filter if power fails.

B J BOSY, *Cambridge, Mass*

5.3 Why does deaerator piping corrode?

Our manufacturing plant is served by two low-pressure (2-psig) deaerators for boiler feedwater, and by a larger, high-pressure (140-psig) deaerator for another plant service. On the low-pressure units, gage glasses usually stain with rust within weeks of cleaning. On the high-pressure unit, we have had severe corrosion of water columns at and above the water level, resulting in several replacements of column piping and level-alarm/control fittings. The materials that are corroding are steel and cast steel. There is some corrosion below the water level, but it's not so severe as at or above the water level. We also have had corrosion in return lines. Present water treatment to the deaerator with the worst corrosion is addition of trisodium phosphate and sodium sulfite, with hydrazine and neutralizing amines injected into the boiler. This treatment started about eight months ago.

Noncondensable gases are vented effectively, we believe. The vent pipe is at the deaerator top, through the center of the internal makeup waterbox. Oxygen content of the water is about 0.400 ppm before deaeration, and about 0.005 ppm after deaeration. Makeup to the closed system, which includes the deaerators, is by process condensate. Have POWER readers had experience with similar difficulties? Can readers tell me why the l-p units' piping corrodes less quickly than the h-p unit's? Should we change water treatment now, or wait and see?—LN

Check the oxygen measurement

Presence of corrosion at and above the waterline of the h-p deaerator unit indicates oxygen corrosion, and this is likely to be the problem in the l-p units, too. Oxygen could come from process leakage or deaerator upsets. Deaerators are stream strippers, designed for specific liquid/gas (feedwater/steam) ratios. If a system has an abrupt increase in liquid-flow rates without corresponding increase in steam rate, the liquid will condense the steam, flood the column, and carry oxygen into the deaerator storage section.

I don't have great faith in the oxygen measurement, because it is presented as a single information point, probably taken under ideal conditions to test the equipment. Also, the chemical and electrical-potential oxygen-measurement methods do have inaccuracies. I concur with the use of sulfite as oxygen scavenger in the deaerator, but I don't believe trisodium phosphate is advisable. It may precipitate with calcium as sludge in the feedwater system. LN should consider raising the l-p units to 10 psig to get better oxygen stripping. Deaerated water for boiler feedwater systems should be stored "cold," with 100-150 ppm $SO_3^=$ (sulfite) to prevent absorption of oxygen in vacuum leakage.

W R NIXON, *Martinez, Calif*

A vent-rate increase could help

A treatment system eight months in operation will do little for a badly corroded piping system, because the chemicals do not restore the missing metal. Ideally, the chemicals would react with sound metal piping surfaces and protect them, but now the best they can do for LN is to slow the corrosion process.

More directly, the rusty glasses indicate presence of iron and/or manganese compounds. Even more likely as a cause is high ferric concentrations picked up by water traveling the corroded piping. The larger process deaerator is getting more severe corrosion because of its higher operating pressure, and consequently higher temperature, about 360F. The smaller units on boiler feedwater probably have some external treatment distinct from that of the larger process unit, and they operate at lower temperature and pressure, enabling them to resist the corrosion process more readily. Possibilities for LN:

■ Check with other plants nearby on nature of chemicals, water source, corrosion problems.

■ Check plant piping for dissimilar metals, especially in the return piping.

■ Revamp condensate-return piping, sloping it for quick drainage, and going to stainless or high-alloy pipe.

■ Line the inside chamber of the larger deaerator with a corrosion-resistant material.

■ Add chemicals upstream of the deaerator.

■ Increase the vent rate, which now may be enough for dissolved oxygen, but not for large amounts of other gases.

■ Dump a percentage of the corrosive condensate, if water supply and costs permit, so more makeup will be fed.

J E HILSON, JR, *Scranton, Penn*

Iron removal may be necessary

Residual oxygen and carbon dioxide are causing most of the corrosion here. Even the reasonable value of 0.005 ppm of dissolved oxygen is corrosive in heated water. Even in the absence of process contamination by leakage into the return condensate, this heat effect will explain the higher corrosion rates in the h-p deaerator. The rust stain comes from iron from corrosion in a large system of heat exchangers and return piping. The iron could be in the makeup water, too. Demineralizers are not highly effective in iron removal, so LN should take water samples from condensate, and also test the makeup water for iron.

Many newer industrial plants with extensive heat-exchange and condensate systems are installing condensate polishers, to protect the boilers from scale resulting from iron and traces of hardness picked up in the condensate. The many cycles of concentration that result from reduced blowdown, where demineralized water is the source of makeup, can build iron to scale-forming levels in the boiler. In mid- and high-pressure range boilers, iron as low as 1 ppm can cause scale-formation problems.

To control heat-exchanger and condensate-line corrosion, feed hydrazine and combined filming and neutralizing amine into the boiler feedwater as far back from the boiler as practicable. This feed point is often above the water level in the deaerator storage space and below the deaerating section. Feeding hydrazine and amines here protects storage space, preboiler equipment, and economizer against corrosion. The location also allows maximum time for the hydrazine to react with dissolved oxygen.

The amine should be fed continuously, at a rate that will maintain a pH of at least 8.5 in the feedwater to the boiler. When this treatment is being started in an existing system, however, apply amine at a low rate, because filming amines tend to loosen and remove old corrosion films. A gradual removal of corrosion products is desired. Hydrazine is fed continuously, to maintain an excess residual of 0.05-0.1 ppm. This level leaves little excess to take care of any oxygen surges or feed-rate upsets. So if the boiler pressure is below 700 psig, a feed of sodium sulfite may be desirable to give an excess residual of 15-25 ppm. A sudden drop in excess sodium sulfite indicates an upset in hydrazine feed or an oxygen surge.

B M KINE, *Vancouver, BC, Canada*

Excess of amines produces carbonic acid

Most probably the amines, if in excess, are causing the carbonates to break down and react with the noncondensable gases (CO_2) to form carbonic acid. This acid will cause excessive corrosion above the waterline in the piping systems, especially where the gases condense. Although not mentioned in the problem, corrosion is probably occurring where condensate lines enter the 1-p deaerators. Solution depends on exact circumstances, but increase the 1-p deaerator water temperature to the maximum allowable, and adjust amines under direction of the chemical supplier's rep. The 1-p units' piping corrodes more rapidly than the h-p unit's, either because feed rates differ or because the temperature of the 1-p systems lets CO_2 condense faster.

J J TANSEY, *Hawthorne, NJ*

Don't inject chemicals into deaerator

LN's problem is most likely CO_2 corrosion or a combination of CO_2 and O_2 corrosion. His recently started water treatment should prevent the corrosion, if control and injection are proper. He should not inject the oxygen scavengers, sulfite and hydrazine, into the deaerator, because they will be neutralized by the O_2 present and will not protect the boiler as intended. Inject the oxygen scavengers into deaerator storage space, or downstream of the deaerator.

The two scavengers are not used together normally, since each does the same job. At elevated temperatures, the hydrazine may decompose to ammonia, go over with the steam, and help neutralize CO_2 downstream. To avoid loss to boiler blowdown, inject neutralizing amines into the steam line at the boiler, and keep the return water at pH of 7.5 to 8.0. At high pressure, more CO_2 or O_2 will be absorbed in condensate, accounting for the higher corrosion rate.

D CRAIG, *Sarnia, Ont, Canada*

Can high-pressure deaerator pressure be cut?

The corrosion indicates overtreatment of the feedwater and corrosion by ammonia. The dissolved-oxygen reading should be 0.003. Morpholine, instead of sodium sulfite, should be used with hydrazine. Does LN really need 140 psig in the deaerator? The elevated temperature causes much faster corrosion, especially with ammonia.

D DUVO, *Northridge, Calif*

5.4 Piping elements for sludge lines

Piping from clarifier bottoms could be a problem in our plant. We have clarifiers on makeup water for the boiler plant, as well as on wastewater cleanup. We must periodically pump out the bottom sludge to a permanent pond. When we installed the clarifiers, we did much of the piping ourselves, and the sludge lines did not receive careful attention. We get abrasive sand and some corrosive liquids in the waste lines, despite care with filters and in the plant processes. Our pH readings are usually 7-9, but have occasionally gone to the acid side—down as far as 4.0.

We have been using steel pipe and iron fittings. On small lines, we installed ball valves. The bigger lines—over 3 in.—have gate valves. Although the systems are only a year old, we are experiencing minor leaks on several valves. We are considering changing to fiberglass-reinforced plastic pipe, but we don't want to do so unless we can be sure of much-improved performance. Can POWER readers give us any advice on this type of piping service? Do expensive alloy valves and special piping materials pay off here, or should we stay with the way in which we have begun?—CRK

Go with neoprene-lined pipe and valves

Our firm has used fiberglass-reinforced plastic pipe successfully with sand slurries. Support for FRP pipe needs special consideration, however. In an existing installation, I would recommend neoprene-lined weir diaphragm valves and neoprene-lined carbon-steel pipe. The lining hardness should be 40 durometer or more. CRK should compare total cost of FRP pipe, which will include additional supports, with the cost of lined pipe, for which the existing support will be adequate.

J A HENSON II, *Lakeland, Fla*

Check space before installing FRP

The chemicals that CRK probably has in his clarifier bottoms should not be a serious threat to his piping. If our experience is close to his, chemicals entering the clarifier should be diluted enough so that the low pH value he mentions will appear only occasionally, and then for only a few minutes. What is probably far worse is the erosive effect of sand and silt, and for this the steel piping will be unsatisfactory over the long run.

One possible source of high cost in replacing existing steel piping with fiberglass-reinforced pipe is lack of space around the present piping runs. This makes it difficult to connect pipe sections by welded joints. We had to take out (temporarily) some of the pipe runs near the steel pipe to be replaced, to give us enough room to install FRP pipe. Hanger spacing is no trouble with FRP pipe. We have mounted replacement FRP pipe with existing hangers and had no trouble. Perhaps our old hangers were too closely spaced, but perhaps the manufacturers of FRP pipe are too cautious. We never hang steel pipe directly from FRP pipe, however.

To keep the cost down, perhaps CRK can quickly check his piping to spot the runs that are more likely to be cut out by sand and silt, and then he can replace a few lengths there. On the lengths near the pond, FRP pipe will be much easier to handle in the necessary changes of location, so perhaps he can start there.

D WHITCOMB, *Savannah, Ga*

Try plastic pipe, fittings, and valves

There are many types of plastic pipe available, as well as pipe fittings and valves. Plastic pipe must be supported at closer intervals than metal pipe, because the beam strength is not as high. Installing abrasion-resistant, corrosion-resistant pipe, fittings, and valves now will save much maintenance. Chemicals to neutralize the corrosives require systems for chemical handling, and these systems may break down and get the plant into difficulties with EPA. Also, filters for filtering abrasives can break down or clog. If metal piping is the weak point in CRK's system, he should go now to plastic pipe, fittings, and valves.

C R KLEIN, *Wauwatosa, Wis*

Limit sludge velocity in piping to 5 ft/sec

The abrasive sand and corrosive liquids in the sludge lines make it advisable to change the existing carbon-steel piping to FRP piping. A recent cost study reveals that the initial installed cost of FRP piping is only about 2% higher in 4-in. diameter and about 5% higher in 6-in. diameter than the cost of carbon-steel piping. The slightly higher installed cost will be offset by the lower maintenance cost compared with carbon-steel. In FRP piping, design so the sludge velocity doesn't exceed 5 ft/sec. Remember that leaks through the valves may be the result of erosion of valve-seat material or improper tightening of packing. Plastic-lined valves might help CRK.

V S R KRISHNAN, *Houston, Tex*

CPVC is least expensive, but support it well

CPVC piping with diaphragm-type valves should give the least-expensive method. Several types of thermoplastic or plastic valves are available, too, and should work here. With CPVC, close attention to piping supports is needed, but a CPVC piping dealer can give CRK the necessary information. If the pressure and temperature ratings for the CPVC piping are out of the range of the clarifier system, FRP piping will serve well. Diaphragm valves can be specified here, too.

R S DUBOVSKY, *Hopewell, Va*

Long-radius elbows help cut wear

CRK's way is going to be more expensive than cheap if he is getting problems with a year-old system. Although the leaks seem to be in the valves, he expresses concern about his steel pipe and cast iron fittings, so he obviously knows that the conditions are going to corrode his steel piping quickly. A fiberglass-reinforced pipe with a wear-resistant liner is suitable for many abrasive fluids at velocities to 6 ft/sec. In one of these pipe types, the 50-mil liner is reinforced to 80% of its thickness with mineral fiber. Long-radius flanged or bonded filament-wound elbows of similar design are also available.

D A Tabel, *Omaha, Neb*

Diaphragm valves should work here

Since velocity is one of the chief causes of wear in piping that handles abrasive materials, perhaps CRK should replace badly eroded sections of pipe with larger-diameter pipe. For example, take out a leaking piece of 3-in. pipe and install 4-in. pipe as a replacement. Velocity would fall to about half the original value, and the replacement would last much longer. This change will cost less than FRP pipe, and there would be the added advantage that joint makeup would be easier.

For throttling needs in handling clarifier sludge, we have obtained good results from diaphragm-type valves in sizes to 3 in. This type seems to resist scouring better than other types, and on the low pressures here the shutoff is reasonably tight. Make sure that a closed valve of this type is really closed, because wiredrawing will cut through the seat in a narrow zone, so the valve cannot close tightly no matter what the closing force.

W G Myers, *Atlanta, Ga*

Get the right type of FRP pipe and liner

Most people don't install special materials on piping such as CRK's clarifier drains. I don't consider FRP pipe as a special material, however, since it sees such wide use. From what the problem indicates, the pumps are holding up, so I infer that they are of recommended material and construction. For piping replacement, FRP is a good choice, but it should be lined with an abrasive-resistant resin and filler. A pipe to carry liquid continuously differs from a vent stack or an occasionally used waste drain. Make sure that the supplier knows what the service is to be.

K Y Lachmann, *Chicago, Ill*

Ball valves with TFE seats serve us well

Ball valves with TFE seats should work on clarifier sludge lines. We have had good results from them, in stainless steel, on lines to 3-in size. We make sure that the valve is either fully open or fully closed, of course. This protection prevents the abrasives in the fluid from tearing out the seat ring if the valve is partly open. The wiping action of the ball past the seat ring seems to keep the silt in the fluid from getting into the valve bonnet. I believe a plug valve would work on this service, too, because it has the same wiping action and close clearances, but we have not had experience with plug valves yet.

G R DAMIANO, *Houston, Tex*

5.5 Improving fuel-oil-strainer operation

Some of the recent problems with fuel oil are more serious for small plants than for large ones. At least, that's what I think after a recent short-term stint in a small plant with a modern 30-gph commercial air-atomizing oil burner on No. 6 oil. Sounds like a simple routine job, but what turned it into a nightmare was the delivery of a 3000-gal load of dirty oil. The burner began cutting out frequently, requiring manual reset each time. The pump-suction strainer had to be cleaned morning and night, so I began to think about ways to improve matters.

In checking over the fuel-handling setup, I found that the equipment was so crowded that I couldn't install a duplex strainer. The suction-strainer basket was of perforated metal, giving good protection but unfortunately clogging rapidly. The easy way out would have been to replace the basket with a coarser-mesh variety, but I was afraid that this would only worsen matters by clogging the pump. I had some other ideas on screens, but I'm still not sure if any minor changes would have worked. Have POWER readers found ways to improve strainer performance on dirty heavy fuel oils? How could a plant engineer judge the merits of such improvements? And how could he specify performance for a new setup?—LAN

Tank bottom might hold answers

This problem is not unusual at all. There are thousands of routine jobs that are pussycats until a load of dirty oil, a bad tank bottom or some high-pour oil arrives to give the recipients a tiger by the tail. I would first install an edge filter with gages on the inlet and outlet sides, and would operate the edge filter whenever any appreciable pressure drop occurs across it, as determined by dual pressure gages. Step two would be to have a competent fuel specialist sample the tank bottom. Step three would be determination of composition of the fouling or dirty material. Put some in a tin dish and heat to about 125F; if it melts down like shoe polish, there is a wax problem. Tank-temperature increase helps with this problem. A good sludge solvent is the last step.

R F SUMM, *New Haven, Conn*

Replace strainer by edge-type filter

LAN's "occasional dirty fuel" problem could likely be solved by substituting an edge-type filter for the existing basket-type suction strainer. Edge-type filters are widely used in lubrication systems of coal-mining equipment. The filters typically have an external handle that scrapes accumulated dirt into a trap for periodic removal. I doubt that LAN's fuel is dirtier than a coal-mining machine's oil, and the viscosities ought to be similar. LAN could probably buy a pressure-drop indicating gage, or even an alarm system to indicate handle-turning time. These filters are exceptionally compact for the volume of oil handled—they have to be, to fit inside low-profile machinery.

D A PATTISON, *Frenchtown, NJ*

Try second screen in fuel-oil strainer

A simple expedient allowed me to remedy a similar problem where coarser-mesh replacement baskets would have passed enough dirt to clog the pump. I made some experiments and finally added an auxiliary screen made of ⅛-in. square-mesh galvanized-wire hardware cloth. This was set inside the basket on the downstream side, and held in place by the flow (sketch, 5.5A). With the

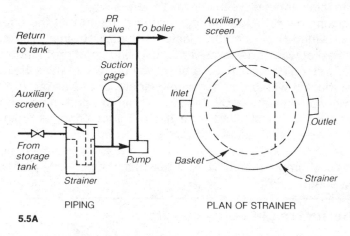

PIPING   PLAN OF STRAINER

5.5A

auxiliary screen installed, the suction gage reads 2 in. higher than before when the screen is clean. The clogging is considerably slowed, and now once a day is enough for cleaning to keep the burner running. The suction gage will read as much as 20 in. as the strainer fills up.

A M PALMER, *Brooklyn, NY*

Install separate pump and filter system

If LAN does not have enough room to install a duplex filter in the system, I would suggest that the best temporary solution is to raise the suction line to a level near the top of the fuel level. The line would have to be lowered every few days to keep it from sucking air, but the higher location would get the line out of the dirtiest oil. At the same time as LAN does this, he should, if possible, install a separate pump and filter to draw from the tank bottom, filter the oil, and return it to the top of the tank (sketch, 5.5B).

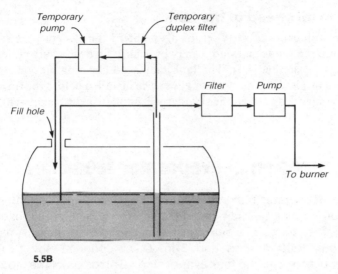

5.5B

Any new filter installation should be of the duplex type, of course, but another worthwhile possibility might be a receiving tank to hold small deliveries until they have been tested and accepted. Any tests that LAN proposes should be discussed with the supplier. I have found that, when approached with the fact that their product is going to be tested, most suppliers are more cautious about what they deliver.

B A WASHINGTON, *Twinsburg, Ohio*

Solvent helps, but watch for water, too

LAN's problem is not with dirty fuel oil. All No. 6 oil may be dirty. The problem is with sludge that accumulates in the bottom of the fuel tank. Sludge is drawn through the suction line and blocks the pump-suction strainer, especially when the oil in the tank is low. LAN should add a fuel-oil solvent to thin the sludge in the tank.

Additives on the market for this purpose will cut the need for cleaning strainers and heaters. Sludge will not accumulate to cause shutdowns, but will be burned in the boiler.

Sludge solvent, similar to kerosene and handled accordingly, is poured into the tank through the fill line. About 1 gal per 2000 gal of No. 6 fuel oil is the ratio for the first treatment. Then add about 1 gal with every 4000 gal of oil after that, or follow instructions of the additive manufacturer. Many plants dispose of waste material, such as combustible solvents and crankcase oil, by dumping them into a fuel-oil tank protected by sludge solvent. The amount of water at the bottom of the fuel-oil tank should be checked, too. Push a measuring stick to the tank bottom, remove the stick, and look for small bubbles on the stick end. Water should be pumped out through a small pipe run down through the fill pipe to tank bottom.

G H HILL, *Sunbury, Penn*

Strain the oil ahead of the tank

LAN blames delivery of dirty oil for his problem. This is not necessarily the case, because the solids and sludge may have built up over a long period, finally reaching the suction inlet. The tank may need cleaning but, once this has been done, a duplex strainer with gages can be installed before the fill opening. A rapid pressure drop across the strainer will indicate a dirty shipment.

J F Woods, *Cockeysville, Md*

5.6 Corrosion in evaporator feed line?

Our boiler-feedwater makeup-treatment system includes filters, weak-acid cation-exchanger, salt-regenerated strong-acid cation exchanger, forced-draft degasifier, steam deaerator, and evaporators. A mild steel pipe, about 1000 ft long and 3-in. O.D., connects the two water-treatment plants with the four evaporators. Corrosion of the pipeline has been a big headache over 15 years. The pipe was replaced with PVC twice, but that, too, failed. Stainless steel is expensive, and mild steel costs nothing initially, because we put in old boiler tubes.

Effluent from the degasifier has about 4 ppm free CO_2, total alkalinity of 0.02 epm, and pH of about 7.0. We have been advised to feed caustic soda after the degasifier to raise pH to 8.7 or higher. We hesitate to do this, however, for fear that the fixed CO_2 will be released in the boiler feedwater and cause trouble. It could also impair steam deaeration at the evaporator. We add cyclohexylamine and hydrazine to the boiler feedwater. Have Power readers experienced similar problems? Should we follow the recommendations for addition of caustic soda after the degasifiers?—MUL

Stainless steel would be best choice

The most-likely cause of the steel-pipe corrosion is oxygen corrosion, with a possible smaller amount caused by residual CO_2 in the degasifier effluent water. The degasifier effluent will be saturated with dissolved oxygen, which is highly corrosive to mild steel. If the water velocity is high through the pipeline, the corrosion will be accelerated by the resultant removal of the protective covering formed by the corrosion product. This will expose fresh surface to the attack. The long pipe run through an operating boiler plant would tend to warm the water in its passage, increasing the corrosion from oxygen. The potential corrosive attack of oxygen doubles for every 30-deg-F increase. The effect will be similar for other dissolved gases, too.

Stainless steel pipe would probably give the best protection against this form of corrosion. Pipe equipped with a corrosion-resistant lining can also be used, but it will likely be more expensive than stainless steel. In condensate systems, the continuous addition of a filming amine inhibitor gives good protection against both oxygen and CO_2 corrosion of mild steel. In MUL's application, however, the advice of a water-treatment specialist should be sought before starting amine treatment, because conditions are somewhat different.

B M KINE, *Mackenzie, BC, Canada*

Lower CO_2 content in degasifier effluent

The severe corrosion experienced by MUL in the mild steel feed lines is primarily caused by the low pH of the feedwater. The presence of 4 ppm free CO_2 further aggravates the condition. Boiler-water pH is typically maintained near 11 to promote formation of magnetite, Fe_3O_4, which stifles corrosion. The pH of 7.0 in MUL's system is far too low to form this barrier layer of magnetite, and severe corrosion is the result. The 4 ppm CO_2 accelerates attack by forming soluble ferrous carbonate, $FeCO_3$.

Although the pH of the degasifier effluent should be raised, caustic soda should not be the selection, because it will fix the free CO_2, which would impede additional removal of CO_2 in the evaporators. If subsequently released in the boilers, the fixed CO_2 would intensify corrosion in boiler tubes, condensers, and return lines. It would also be recycled indefinitely in the system until blown down. Ideally, the CO_2 content of the degasifier effluent should be lowered, after which the effluent pH could be safely raised by caustic soda—or preferably by a coordinated phosphate treatment. If a lower CO_2 content is not feasible, the safest solution would be to raise effluent pH to about 11 but to acidify to 7 in the evaporators to release the CO_2 fixed by the caustic soda or phosphate additions.

R B DIEGLE, *Marion, Ohio*

Look into galvanic action, too

It sounds like galvanic action to me. How many dissimilar metals are in all the piping? Insulated unions between pipe of different metals or units containing different metals are easy to install and could settle the matter.

G ROSEKILLY, *San Mateo, Calif*

Stainless—and check size needs

MUL has his cost priorities in disorder, in my opinion. This looks to me like a simple summing up of total annual costs for one pipe type vs another. To start with, MUL is entirely wrong when he says mild steel costs nothing initially because he uses old boiler tubes. He forgets that the costs of inspecting, cutting, preparing for welding, welding, and hanging are very real. He should not forget, either, that fittings for boiler tube are hard to find, unlike the common screwed or flanged fittings for regular pipe.

On the other hand, if he buys thin-wall stainless pipe he can get far better corrosion resistance. He might be able to go to a smaller size of pipe, too, since apparently the availability of 3-in. boiler tube, rather than the hydraulic-flow need for that size, was what prompted its use originally. If the stainless lasts far longer than the mild steel, MUL will be ahead in total annual costs, in addition to avoiding extra shutdowns and loss of treated water.

A W Le Brun, *Rochester, NY*

FRP pipe has worked on condensate

Instead of adding a variety of chemicals to control corrosion in the feed line, MUL should install fiberglass-reinforced plastic (FRP) pipe to solve his corrosion problem. We have used FRP pipe in 2- and 3-in. sizes for over six years and have had no corrosion failures in steam condensate systems. We put filament-wound fittings in these lines. Pressure/temperature ratings for various sizes and for fittings are available from the manufacturers.

D W Antonacci, *Cocoa Beach, Fla*

Caustic soda is cheapest solution

Adding caustic soda after the degasifier is the most economic solution to MUL's problem. His fear that the fixed CO_2 will be released in the boiler feedwater appears to be unfounded. When the pH of the treated water is raised, the carbon dioxide in solution as carbonic acid (H_2CO_3) will become ionized as bicarbonates and carbonates:

$$H_2CO_3 \rightarrow H^+ + HCO_3^-, \text{ and}$$
$$HCO_3 \rightarrow H^+ + CO_3^{--}$$

In these ionized forms, the amount of CO_2 carrying over from the evaporator to the feedwater will be reduced. What CO_2 does come off will be removed by the deaerator in the normal way. The present use of cyclohexylamine as a neutralizing amine should offer sufficient protection against any CO_2 that manages to find its way into the boiler feedwater.

If MUL still wants to avoid caustic soda, then a second solution to the problem would be rubber-lined steel pipe. Although initial cost would be high, it would still be only a fraction of that for stainless steel. Our plant has been handling low-pH (4-6) treated water for the past 20 years with rubber-lined pipe, and we have had very few problems.

J J Mucci, *Bronx, NY*

Try PVC again, but watch specs

The cyclohexylamine is not going to do MUL any good in this line, and the hydrazine is not added until later, in the boiler. Either more chemicals are wanted, or a close look at piping materials is in order. I think that the cost of constant chemical addition will be found higher than proper materials for the line.

PVC should be adequate, if quality is right and assembly is as specified. Not all PVC piping meets the standards claimed, so MUL might check around to verify that his PVC pipe is on-spec.

J R Walton, *Baton Rouge, La*

5.7 Removing oil from return condensate

Our plant needs a method for removing oil from return condensate. We have steam-driven compressors, which are supplied with steam at 400 psig, 550F and which discharge into surface condensers at 24 in. Hg vacuum. At present, we inject a floc solution prepared by combining soda ash and filter alum into the dirty condensate. The condensate is then pumped through oil-removal filters having classified anthracite coal as filter bed. These filters work fairly well, but need to be backwashed periodically. The backwash effluent goes into the plant outfall, and this is what has caused our problem. The oil content of the dirty condensate is 125 ppm free oil and 25 ppm emulsified oil.

We need an oil-removal system that will remove free oil and emulsified oil from the dirty condensate and transfer these oils into a bulk container for haul-off. The clean condensate from the system should have an oil level below 1 ppm to allow us to put it back into our boilers. Flow: 300 gpm max. Have POWER readers had experience with such a system, and is the cost reasonable for our type of operation? Will centrifuging work on this service, or are there more efficient oil-removal filters than we are using? Can we store the backwash and release it a little at a time to reduce maximum values of oil in water?—VO.

Put in an oil separator ahead of filters

For a problem similar to VO's, we evaluated two types of mechanical separators to see if free and emulsified oil could be removed satisfactorily from oily condensate. One unit was a coalescing type; the other removed free oil on inclined baffles, according to Stokes' law, with the emulsion broken by impingement and preferential wetting. Neither unit successfully reduced oil content in the effluent to below 2 ppm.

Now, because of the urgent need to reduce the amount of oil in plant effluent, we are going to buy an oil separator and install it in front of oil-removal filters. The separator should take out about 95% of the oil in the dirty condensate, and the oil-removal filters will be expected to remove the rest. We will backwash the filters, but frequency of backwashing should be materially reduced. We would be interested in a system to reduce oil to below 1 ppm in the condensate with no backwashing of filters. Perhaps this is asking too much, and perhaps water-quality boards will modify their requirements. "Zero discharge" makes me shudder when I think of it.

C VINCENT, JR, *Orange, Tex*

Risk outweighs gains: dump condensate

Common engineering practice is to waste such condensate. Considering the small economy obtained, cleanup is not worth the risk. A little oil in the wrong place can do immeasurable damage to high-temperature boiler tubes. Ecological considerations may dictate separation of oil from the contaminated condensate by settlement or other means, but the separated water will not be returned to the circulating condensate.

A M PALMER, *Brooklyn, NY*

Replace spent anthracite—it's cheaper

Yes, the job can be done, but why go to the trouble and expense? The anthracite coal that has done its work as a filtering medium should be discarded and replaced by fresh, clean anthracite coal, which is cheap. One filter bed in use and one on standby will allow a bed to be removed and replaced.

E KAKNICS, *Philadelphia, Penn*

Chill the oil to 34F to separate it

Perhaps the backwash effluent could be filtered through anthracite, then chilled to 34F by passage across refrigerant-cooled plates, with the water dripping down and the solidified oil removed by conveyor and reheated to 180F.

D F INGCO, *East Orange, NJ*

Get 2.5 ppm with coalescing unit

A coalescing unit with polypropylene elements is removing the same type of lubricant from steam condensate in a similar operation. Inlet oil concentration of 100-200 ppm is reduced to 2.5 ppm in a two-stage separator. Pressure drop varies from 2 psi at startup to 30 psi when the elements become loaded with solids. Depending on cleanliness of the condensate, some prefiltration may be necessary. No chemical additives are required. One advantage of the method is that it is entirely mechanical. Boiler-feed chemical consumption is reduced by the reduction in makeup water, too, and the energy requirements are less because the unit operates at 180-200F. No backwash is necessary in a separator of this type.

C W DAVIS, *Roxboro, NC*

Single-effect evaporation could do the job

Perhaps VO's problem might best be solved by a method not often considered for waste waters—evaporation. His present system is satisfactory except for the problem of backwash-water disposal. Since the volume of backwash is undoubtedly only a small fraction of the volume of condensate treated, he has already taken a big step toward reducing the volume needing treatment. All he now requires is a holding basin for his backwash, and a small evaporator to concentrate waste.

Single-effect evaporation is probably all that can be justified by the volume. The evaporator could be located near the steam engines, so the vapor could be condensed by the engines' condensers. This would also provide some distilled boiler makeup, and any oil steam-distilled would be taken out by the present filters. A forced-circulation evaporator would probably be best, both to reduce fouling and to permit concentration to nearly water-free state. The evaporator vapor head can usually be made large enough to allow semibatch operation. Feeding would be continuous, with concentration building up until, at the end of a week, the evaporator would have enough oily residue to fill a tank truck.

Instrumenting the evaporator for untended operation should be easy, as shown in 5.7A. Vacuum provided by the existing condenser draws in the impure feedwater. The only pump is for forced circulation through the heater, and it can also transfer evaporator contents to disposal. Feed rate is controlled by evaporator level, with an override to prevent sucking air from an empty sump.

Steam is manually controlled, at a rate sufficient to handle the backwash volume in the time available, but with an override to prevent overheating when all water has been removed.

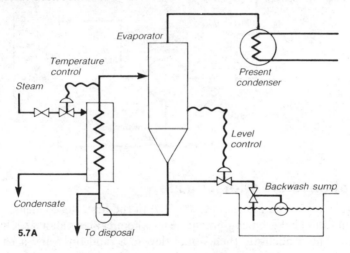

If VO's waste volume were large enough, or his heat supply limited, the multiple-effect principle might serve, with operation still unattended. We have such systems handling all wastes from a dyestuffs plant, and oily wastes from parts washers in a metal-finishing plant. Although energy costs for these systems are higher than for some alternatives, such as flotation, first cost is usually lower.

F C STANDIFORD, *Ann Arbor, Mich*

Consider centrifugal extractor

As in many phases of water cleanup, oil-separation systems become more expensive as the last bit of oil is removed. A gravity separation system, consisting of a large tank baffled so that the contaminated condensate must make a long journey through the storage, will allow most of the oil and emulsion to rise to the surface. There it can be skimmed off or picked up by an absorbent belt, after which the flotate can be sent through a more active separation device for final cleaning.

Some very efficient centrifugal extractors are available for this service (sketch, 5.7B). They are useful in handling liquids of low density difference and those with emulsion-forming tendencies. Another means consists of a device with coalescing and settling membranes to separate the fluids. Some of the devices are rated to 1100 gpm, much more than is needed here.

B B Brown, *Hagerstown, Md*

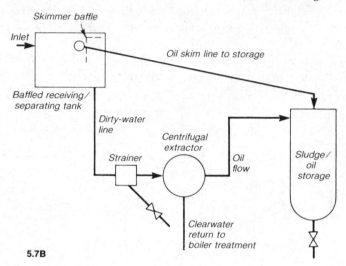

5.7B

Adsorption filters offer long life

There are two possible alternatives for VO. In the first, the condensate would be treated in flocculation equipment, using chemical coagulants such as lime and alum. The condensate then would flow to a sand filter and a battery of adsorption filters with activated carbon. Condensate from the adsorption filters would contain 1-2 ppm organic matter. Regeneration of the activated carbon is by washing with gasoline at 176-140F. Gasoline remaining in the filter bed is eliminated by steaming. Lifetime of the activated carbon is long. The alternative would be to wash the existing anthracite filters with gasoline, and to remove residual gasoline by steaming. Cleaning the gasoline is done in a distillation plant.

B Danilov, *Haifa, Israel*

5.8 Is phosphate the best boiler treatment?

Our two high-temperature hot-water (HTHW) boilers operate without a water softener to precondition water. The system is closed, requiring little makeup. The boilers had been treated with chromates until we were forbidden to use them a few years ago. For nearly a year, the HTHW boilers received no chemical treatment. We then obtained permission to treat with phosphate, with instructions to keep the ppm count at about 20. We were assured by the chemical vendor that we could maintain this level with periodic testing and correction. After a few months, a change was recommended, but I got the vendor to agree to test the water monthly and make the necessary chemical adjustments. Our ppm readings have ranged from the present 280 to 1200. Our state's Dept of Health has verified this. (The vendor insists it is only 20-30 ppm.)

Following this, we have had several vendors approach us to recommend various treatments. There is very little information available on HTHW treatment, and I wonder if there is any general treatment that is good for all areas of the county. Can POWER readers tell us if phosphate treatment is considered good for HTHW boilers? If not, what are the reasons, and what is a good treatment? Can I write a set of specs that I can give to every chemical vendor, with assurance that, if he meets them, the treatment will be effective?—BID

Treatment and testing—both are musts

Water is not identical in chemistry in all parts of the country. For this reason, no general treatment is possible that will be good for all areas. As far as phosphate is concerned, we are keeping phosphate in our boilers from as low as 1 ppm up to 30 ppm, depending on the type of unit. Among other chemicals, we use disodium phosphate and trisodium phosphate prepared with sodium nitrate. Our steam generators demand rigid water quality; thus, testing and treatment is conducted at least every other day, and in some cases is automatic. BID did not tell the size of his plant. If he can afford it, however, a biweekly test and treatment program seems in order. Unless BID or somebody on his staff is well versed in boiler-water chemistry, his best bet is to hire a consultant to determine plant needs.

E C CUBARRUBIA, *Avon Lake, Ohio*

Treatment depends on temperature

By itself, phosphate is not a good corrosion inhibitor. Phosphate will form a precipitate with the hardness in the HTHW system. If you try to remove the precipitate by blowdown, more makeup water must be brought in, leading to more precipitation. As more makeup water is used, dissolved oxygen and carbon dioxide are introduced; both are corrosive. Below 350F, a good treatment for HTHW systems is nitrite borate (500 ppm to 1000 ppm) and liquid caustic soda to maintain a pH of 8 to 10. These materials are not pollutants, but you may have to dilute bleedoff (which is more concentrated) to the municipal sewer system, using its water to meet local disposal limits.

Above 400F, I would recommend using sodium sulfite at 20 ppm to 40 ppm and liquid caustic soda to maintain a pH of 9 or 10. Keep in mind that specifications are only as good as the maintenance behind them. The cost of the chemicals needed to maintain your system would be relatively small; therefore, I would include a service contract in your specification that could call for 12 visits annually, with in-plant testing, written reports and a written guarantee against all corrosion. This will give you the protection you need.

J KORTEKAMP, *Monson, Mass*

Two primary problems: corrosion, scaling

At our company, we do not believe that phosphate is the most suitable HTHW boiler treatment. An alkaline chromate treatment used in the past has provided effective corrosion control; usually, the only protection against fouling deposits was provided by the alkalinity introduced with chromate treatment. It is possible that a completely formulated phosphate treatment may be acceptable for HTHW boilers, but recent advances in water chemistry can provide a more-effective treatment program. The considerable variation in BID's phosphate readings may be indicative of appreciable water loss, and makeup not synchronized with chemical feed. The discrepancy in readings should be resolved, and may be partially due to the type of test made. For example, if the vendor is running a simple phosphate test while the plant and state Dept of Health are determining total phosphates, this could partially account for the discrepancy.

Water treatment for HTHW boilers poses two primary problems: (1) Corrosion, which can lead to system deterioration; and (2) fouling or scaling, which can retard heat transfer and restrict fluid flow. For corrosion prevention, alkaline buffers coupled with oxygen scavengers or corrosion inhibitors are usually selected. A buffered oxygen-scavenger system can be a problem in HTHW boiler treatment, however, because air infiltration may rapidly deplete oxygen scavengers. Many corrosion-inhibitor systems are available; the choice should be at the discretion of BID's water consultant.

For scale control, we believe chelating agents are the best choice for HTHW boiler treatment. Rather than precipitating hardness, which is the case with phosphate or alkaline treatment, fouling materials are kept in solution. This is important because the high circulation rates in HTHW systems do not normally allow precipitated materials to settle and concentrate so they can be removed. Chelating agents have the added value of cleaning all fouling deposits from the system, thus assuring maximum heat transfer and efficiency. Periodic tests for excess chelating agent—and possibly for hardness and iron—would permit effective control of the treatment program. BID would be assured that his system is not being fouled, and is also being protected against corrosion.

G A RESNIK, *South Bend, Ind*

Use borate nitrite, plus testing to record loss

BID is correct in saying there is little agreement among water-treatment vendors on treating HTHW systems. I've been treating these systems with borate nitrite for nearly 15 years with good results, using steel and copper test strips to record the corrosion loss. It protects by forming a strong iron-oxide/borate-nitrite film on the metal surfaces. A leading vendor ran a successful study on this compound at Colorado College, while I was in charge of the plant there. We maintain 2000 ppm nitrite in the system.

Our operating temperatures do not exceed 360F, and we have not experienced any difficulty with mechanical seals on pumps using tungsten carbide material. We do soften the water used for makeup, and BID may have to go this route if a reliable water-treatment company suggests it. As he says, his loss is small, so it would not require a large softener to do the trick. With softened water, he'll skirt many problems.

P McCann, *Santa Cruz, Calif*

Polyphosphates effective only to 180F

Use of phosphate in treatment of HTHW water for scale control is not a good practice. Polyphosphates up to 20 ppm are excellent scale inhibitors only at temperatures up to 180F. At this level, they hold calcium carbonate in solution by the "threshold effect." Above 20 ppm, polyphosphates can sequester hardness when used in proportion to the hardness present, thereby preventing scale formation.

At temperatures above 180F, however, polyphosphates lose their effectiveness, due to reversion to the ortho form. Upon reversion, calcium will combine with the orthophosphate, forming the insoluble calcium phosphate. This can lead to deposits of calcium phosphate on heat-transfer surfaces, thereby contributing to scale formation. Treatment of HTHW systems should include the following:

■ For corrosion and pitting control, maintain pH values of 8 to 9. Use an oxygen scavenger such as sodium sulfite to maintain a sulfite residual at 30 ppm to 50 ppm, or use hydrazine to maintain a hydrazine residual at 3 ppm to 5 ppm.

■ For scale and deposit control, soften the makeup water with zeolite, and introduce thermally stable scale inhibitors and dispersants, such as polyacrylates at 10 ppm to 20 ppm.

Proprietary inhibitor systems incorporating both these control mechanisms are available from vendors.

R T Blake, *Long Island City, NY*

Plotting scavenger gradient will help

Most treatment formulations use surface-active agents, such as phosphate and chromate, to protect against corrosion and scale. Many are fed without proper control limits or control of pH. In many cases, alkalinity, pH, and total solids build up to damaging levels, causing leaks at packing glands, and dezincification of brass components in the system. If the system could be kept tight, no

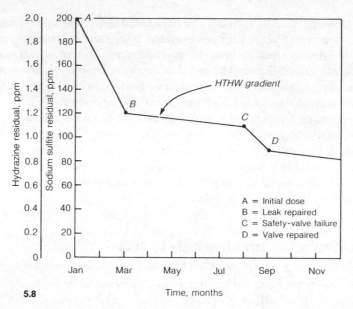

5.8   Time, months

treatment would be necessary, because a state of equilibrium would quickly be reached. Since this isn't possible, mechanical problems occur, leading to water loss and/or air inspiration.

The simplest, least-expensive treatment method is to introduce an oxygen scavenger into the system and plot the gradient of its decline from a specific point in time to another (graph, 5.8). The gradient will determine the relative tightness of the system and will provide a positive means of monitoring system tightness on a routine basis. Improving tightness would be reflected by a leveling of the gradient. Mechanical problems causing loss of water or air inspiration would be reflected by a plunging gradient. The oxygen scavenger (sodium sulfite or hydrazine) will also assure that corrosive oxygen will be eliminated from the system. Adjustment of pH to 9.0-9.9 is compatible with both ferrous and nonferrous components of the system.

U E DISHART, *Pittsburgh, Penn*

5.9 What did spray-nozzle deposits mean?

Two steam generators, each 200,000 lb/hr maximum at 1200 psig and 887F, supply turbine/generator and process steam for our refinery. A few months after startup, the desuperheating spray system gave us trouble— poor temperature control and instability. (We had originally thought the system was too large.) When we looked at the spray-nozzle water holes, we found them plugged by a fine black powder. The nozzle, which resembled a venturi, contracted to a short throat, and then enlarged again in conical form. The water-spray hole openings, slanting downstream, were in the nozzle wall, about one-third of the way along the conical enlarging section. Downstream from them was another deposit of black powder along the conical wall. The powder was analyzed as 51.5% Fe_3O_4, 43.4% Fe_2O_3, 1.5% SiO_2, 1.5% CaO, 0.5% MgO. It was similar to a loose deposit we had found in the lower and upper boiler drums previously, when inspection showed the interior surfaces to be in good condition.

We replaced the nozzles, but they soon replugged, so we changed to a new design. The outer downstream end is now rounded, and the water-spray holes are radial and at the narrow throat. The plugging trouble has stopped, but I still wonder where the powder came from, and whether we should be on the lookout for more difficulty. We hold pH at 11.5 or a little below. Residual oxygen at boiler inlet is under 0.007 ppm, and we demineralize and treat with phosphate, caustic and hydrazine. Have POWER readers experienced similar deposits in desuperheater spray nozzles? Do the deposits indicate trouble elsewhere, perhaps in the boiler or superheater? And what precautions should we take in operation and maintenance?—KT

Treatment chemicals are cause

In general, desuperheater spray nozzles are most likely to be the first elements to show accumulation of chemical deposits from feedwater. The best desuperheater water is, of course, steam condensate—if available. On the black powder, my experience indicates that it comes from oxygen-absorbing chemicals used for feedwater treatment.

A J HERMANS, *Pittsburgh, Penn*

Amine feed is helpful after pH sampling

KT's problem is not with the steam, but with the superheater spray water. We have noticed a similar condition, although on a much smaller scale. We have had no plugging of desuperheater spray nozzles, and the deposits are minor. At our plant, the boiler feed pumps supply both boiler feed and desuperheater spray water. Small amounts of iron oxide and other impurities in the feedwater tend to be left behind in the desuperheater assembly when the spray water is flashed.

Because the same water serves for both boiler feed and desuperheater spray, boiler drum sludge and desuperheater deposits show about the same analysis. Our largest source of impurities in the boiler-feed and spray-water systems is in the plant's condensate-return system from process, which is quite extensive. We not only encounter some hardness contamination from time to time, but we also find that pH control in all distant areas of the plant can be difficult. Where low pH conditions exist, corrosion can be rapid. If the corrosion products go back to the boiler plant in the process returns and enter the feedwater and desuperheater spray, the result is boiler sludge and desuperheater deposits. Taking pH samples at remote points and feeding amines are steps to cope with the problem.

Boiler carryover to the desuperheater is a serious matter, requiring constant attention. Taking condensed steam samples at the superheater inlet is valuable in determining levels of carryover. We think, however, that the source of our deposits is external to the boiler—corrosion products and other impurities carried in the desuperheater spray water.

V SHIRE, *Omaha, Neb*

We vented CO₂ from condenser

KT's problem strongly resembles difficulties that we encountered in first operation of our 825,000-lb/hr, 790-psig, 750F steam generators. Each steam generator has a desuperheating condenser, with a sidestream of boiler feedwater for cooling. The condensate produced in this way is injected into the steam in a venturi-type spray nozzle. Water nozzles in the spray header kept plugging up with black powder like KT's.

A temporary solution was the installation of a blowoff connection that let us clean the water nozzles periodically with steam. This kept us going, but did not eliminate the problem. The composition of deposits suggested presence of carbon dioxide. We concluded that it got into the system chiefly as a product of thermal decomposition of carbonate and bicarbonate alkalinity. Continuous venting of CO_2 from the desuperheating condenser solved the problem.

At the same time, we were experiencing high Fe levels in condensate returning from process. We installed vents on all the process heat exchangers, and also sent all process condensate through the polishing filters. Inspections since have shown no deposits.

V HOJDA, *Ft McMurray, Alberta, Canada*

Nozzle change helped plant in Thailand

We observed plugging and deposits in superheater spray nozzles like those of

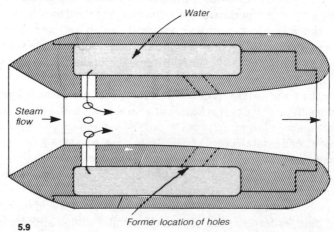

Water

Steam flow →

5.9   Former location of holes

the problem. We thought that the deposits, which were mostly Fe_3O_4 and Fe_2O_3, might be caused by erosion in the superheater tubes because of water droplets carried over. Either high water level in the drum or poor design of the steam separator could cause this. Also, the rough surface of the nozzle body might have caused a partial vacuum behind the downstream face of the square nozzle block, resulting in a deposit of material because of eddies in the flow.

We designed two new nozzles, one with water outlet holes in the diverging part of the nozzle, and one with holes in the throat, as shown in sketch. The plugging has not recurred, and recent inspection showed both new designs to be satisfactory. Stability of steam temperature is much better now, too. A nozzle with a single hole has been tried, but gives poor temperature control.

K K JAO, *Bangkok, Thailand*

Suspect condensate and makeup

KT does not indicate the source of the desuperheating water. Normally it would be boiler feedwater, and if this is the case here, then the deposits originated in the condensate and/or makeup water. If the demineralizer has a strong-base anion resin and produces good water, then condensate contamination and corrosion of the condensate system are the cause.

Deposition in this area does not normally indicate potential problems in the boiler or superheater, unless corrosion products from the condensate system are excessive. The boiler would then get dirtier faster, requiring more frequent chemical cleaning to prevent gouging of the metal under tube deposits. The condensate system should be treated, probably with a filming amine, to cut corrosion to a minimum.

R S MEDVE, *Silver Spring, Md*

Don't forget to clean out nozzle lines, too

The rust collecting in the spray nozzle came from rust in the water side or steam side. A thorough cleaning of the steam drum is advisable at the next plant shutdown. In addition, the water line to the superheater pipe should be disconnected and struck with a hammer, to jar loose any rust stuck to the inside of the pipe. The superheater line and water line to the spray nozzles should then be disconnected, and the loose rust shaken out or blown out with compressed air. Stainless steel would be the best material for the spray nozzles.

J RAMPOLLA, *New York, NY*

5.10 Dewpoint corrosion in i-d fans

Our steam plant burns oil in two boilers equipped with tubular air heaters. These boilers were originally fired by pulverized coal. For several years we burned high-sulfur oil but avoided corrosion damage in heaters and fans by keeping temperatures well above the dew point at heater outlets. Recently we had to switch to low-sulfur oil (at higher prices, of course) to meet emission requirements. The reduced SO_2 concentration in the flue gas now has made us aware of the possibilities for cost saving by decrease in flue-gas temperature. We would like to come down as close to the dewpoint as possible, even at risk of passing below it on occasional low loads.

Short-term tests, while not conclusive on corrosion rates, have indicated that we will get some corrosion. We are concerned more about possible corrosion effect on the induced-draft fans than on the air heaters. Under consideration now are coating of plastics and metals. We are not sure of the benefits or life of these coatings, however, and would like an assist from POWER readers. Specifically, have POWER readers had good results with metal overlays, or have cracks in base metal and breakouts been troublesome? We want to avoid serious vibration problems at all costs, because our fans are not too well mounted. Secondly, how have plastic coatings stood up, and is care needed in applying them to fan rotors?—ARH

Watch for deposits on fan coatings

ARH's difficulties are being experienced by many others at present. One major obstacle to economic solution of the corrosion problem is the size of induced-draft fans and ducting in their vicinity. If the equipment were much smaller, noncorrosive materials could be a good solution, but because of the expense, the tendency has been to stay with ordinary steel and try to coat it. Thick coating is suitable for ducting and breeching, but not for high-speed fans, where a defect in the coating can cause a piece to be flung off and expose the metal to corrosion. Some coatings build up with deposits more than the original metal, so the fan can slowly become unbalanced. For these reasons, probably the old remedy of keeping the inlet temperature 20-deg F to 50-deg F above the dewpoint is the best. In the air-heater sections, the glass tubes seem to have proven successful in several cases.

A CLARKE, *Detroit, Mich*

MgO slurry demands low excess air

Many methods have been attempted to prevent dewpoint corrosion in induced-draft fans and air heaters. Plastic coatings tend to crack and are difficult to apply; pinholes allow serious corrosion below the coating. Glass tubular air heaters are expensive and can crack, as can the ceramic coating on some tubes. The tubes must be installed during outage or overhaul.

Steam loops added to the tubular air heaters to raise air temperature there have been troubled by leaks. Added cost of steam is a factor, too. Injections of ammonia or caustic have formed water-soluble and tenacious deposits which can badly imbalance the fans. MgO slurry additives in the fuel oil can reduce

SO_3 corrosion only if excess air is below the 3-5% range. With higher excess air, the products increase SO_3 concentration and dewpoint, in addition to building deposits.   I KUKIN, *Clifton, NJ*

Clean the fan blades before coating

If ARH decides to use a coating on his induced-draft fan blades, care must be taken in application. Contaminants must be carefully removed before coating is deposited. These contaminants are of two kinds, visible and invisible. The visible ones, grease, oil and dirt, cause much less trouble than invisible ones, such as alkali, salts, acids and finger marks, which when present under a plastic coating, accelerate corrosion and blistering.

Before rinsing the blades prior to coating, protect against alkali in the rinse water by maintaining the solution at pH below 5 with a mixture of chromic and phosphoric acids. Also, remember that compromise is necessary: to gain flexibility, some hardness must be sacrificed; to gain durability, some transparency must be sacrificed; and to gain adhesion, some chemical resistance must be sacrificed.   R G PHEIL, *Racine, Wis*

Added fan warms air at part loads

A better method might be to install an air-recirculation system, as shown in sketch. These recirculation systems are used on some boilers that carry low

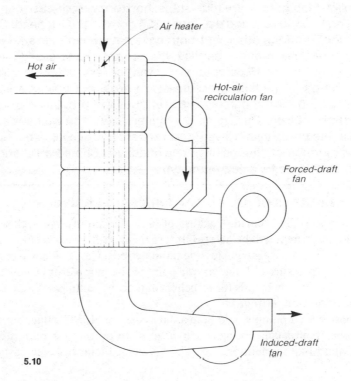

Air heater

Hot air

Hot-air
recirculation fan

Forced-draft
fan

Induced-draft
fan

5.10

loads on occasion. The purpose is to keep the temperature at least 20 deg F above dewpoint. Painting the rotor with a good enamel is also a method of protection. Another protection against corrosion is the use of an alkaline powder, such as dolomite, magnesite, or lime, fired with the fuel and equal in amount to the ash content of the fuel. The alkaline powder helps neutralize H_2SO_4 and stops corrosion, in addition to reducing deposits of particles.

B Baranow, *Plainfield, NJ*

Galvanizing could guard impeller

One solution would be to remove the impeller from the housing, clean it, examine it and, if it does not have to be repaired or replaced, have it galvanized. This should prevent or retard corrosion. Galvanizing produces a chemically inert surface which can stand the temperature of the flue gas without melting. If the fan speed is moderate, vibration should not be a problem. Examination of bearing condition during dismantling will show whether vibration damage has occurred.

E Kaknics, *Philadelphia, Penn*

5.11   Causes of erosion on casing flange

Our boiler-feed pumps are four-stage, horizontal, split-case centrifugals operating at 3550 rpm and delivering 300 gpm at 1700-ft head. One is six years older than the other, but both have developed the same problem: After two or three months' service, their capacity decreased—to as low as 200 gpm for one. Examination showed severe water washing at the casing flange in the high-pressure-stage area, but nowhere else. Both pumps have been repeatedly repaired, but the problem has recurred. Casings are of iron. Our water is demineralized, and we have adequate NPSH at the pump inlet. Have Power readers had similar experience, and if so, what repair or change has been effective? Should we change materials, or should we try a new pump type?—SPS

Check flatness of joint between casing halves

Cavitation may be occuring because of high suction lift. A check on suction piping arrangement will be helpful. Internal cavitation may occur at reduction in maximum capacity during low-demand periods. Cavitation will result when the absolute pressure of liquid at *any* point in the pump drops below the vapor pressure. Special materials for impellers and other pump parts may be needed to resist cavitation pitting.

A booster feed pump would help assure positive NPSH value, but there are some other points concerning overhaul prior to putting the pumps in service that I would like to mention, too. The horizontal joint between casing halves should be as flat as possible. Check to see if the halves rock slightly when pressed against. The casing dowels must be properly aligned with respect to bore center, too. Casing gasket must be proper thickness. When the rotor is installed, the upper casing should be torqued down to squeeze the casing rings slightly .and eliminate seepage and cavitation. Also, the bore for the rings should be checked for ellipticity when casing halves are bolted together with the gasket in place.

C J Myers, *Cape May, NJ*

Flange boltup may offer clue

It would be helpful to see the other half of the pump casing, but it is a fair guess that it has similar erosion channels caused by the interstage leakage that produced the marking shown. SPS should ask the following questions:

- Has gasket material been changed from the original?
- Are flange bolts pulled down in proper sequence in reassembly, or do mechanics draw them down haphazardly?
- Are the flange bolts torque-wrenched or merely pulled to the mechanics' maximum strength?
- Are the flange bolts pulled up again after the pump has reached operating temperature and pressure?
- Have the flange faces been checked with a precision straightedge to see if convexity or concavity exists?
- Is the boiler feedwater deaerated properly?
- Does the pump manufacturer have any solutions from his experience?
- Are records kept to discover just when the problem begins and how long it goes before it becomes intolerable, on the basis of head loss vs time?

It seems a little unusual to have cast iron pumps in 700-psig-plus service. Cast iron is a good material, but steel would be more appropriate in this high-pressure service.　　　　　　　　　　　　　　　　　B B Brown, *Hagerstown, Md*

Oversized pump might be cause

I believe the trouble is an excess interstage pressure differential that is causing a breakdown of the gasket. The resulting backflow to the previous stage erodes the casing metal. High-pressure heated water will flash to steam on release of pressure, and this will cause erosion. Possible causes? Is the right gasket material being used? Is it being installed properly? Is the output pressure at the pump discharge nozzle correct and within design limits? Perhaps the boiler-feedwater regulating valve is sized too small, causing a pressure backup at the pump. An oversized pump, too, could develop an overpressure at the final stage.　　　　　　　　　　　　　　　　　　A M Palmer, *Brooklyn, NY*

Cast iron? Specify other metals

If SPS's pump is of cast iron, it cannot be welded readily, and other materials should be specified when purchasing another pump. The original manufacturer of the pump should be called in to offer an explanation and remedy for the problem. But in the meantime, the flow conditions at the outlet side of the pump should be looked at. This involves the shape of the nozzle to determine how easily water can leave the pump. If there are restrictions to water flow here, the tendency will be for the water to wash away the metal in its path. The flow characteristics are another point to take up with the pump manufacturer. This will improve matters for other customers, too, sparing them the repair cost and downtime.

E L KAKNICS, *Philadelphia, Penn*

Poorly fitting gasket was culprit

Many years ago I had a similar problem with a multistage high-pressure (860-psig) boiler-feed pump. It was eroding between stages. We found that a high

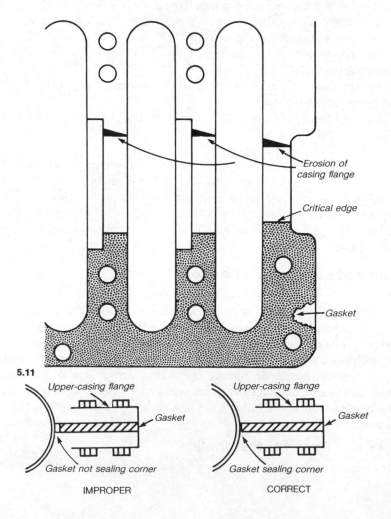

5.11

differential pressure (about 215 psig) between stages, coupled with a poorly fitting gasket, would start a small leak that would erode away the gasket, and then gradually erode the pump-casing flange (sketch). The section with the highest ΔP is most susceptible to this damage. The cure was a gasket slightly oversized at these corners; this makes sure that the corner is tightly filled by the gasket. A high-temperature sealant should also be applied.

F S MARKEN, *Northport, NY*

Look at photomicrographs of worn area

To find a common denominator for both pumps, I would suggest the possibility of cavitation rather than water wash. If SPS were to look at 50X or 100X photomicrographs of the worn area, he could tell the difference. Erosion is obvious from the "cutting" type of wear pattern. In the case of cavitation, the implosions will make the worn surface look as if a little man with a ball-peen hammer had been working on it. The solution? Be sure to eliminate any air leakage and, more important, any signs of starving the suction side.

V L MIDDLETON, *East Alton, Ill*

Is pump capacity excessive?

The operating parameters given by SPS show the pump specific speed to be about 230, which is very low for centrifugal pumps. This will give poor efficiency, especially when operating at below design capacity. In SPS's pumps, the casing joints seem subject to wire drawing. This often occurs when water bypasses the casing rings or diaphragms; water-entrained vapor and hydraulic shock are the cause. Excessive throttling because of oversized pumps or sticky regulator valves are basic reasons. If regulator throttling is excessive, smaller-capacity pumps should be installed. Two half-capacity units would give more efficient operation during low-demand periods.

H B WAYNE, *Jamaica, NY*

5.12 Why sand in filtered water supply?

We pump 4-million to 5-million gpd of water from Lake Erie for use in our mill operations. First stage of water treatment is an Infilco Accelator (capacity 5-million gpd) to clarify the water, which next goes through five sand filters and into our clearwell. Pumps draw the water from a sump in the clearwell and deliver it to the mill at 70-80 psig. Sand is a problem in this system—we find sand in our clearwell, pipe lines, filters, and other places where we don't want it. Our air-compressor heads, for instance, have lost much metal from sand erosion.

Our five filters have received considerable attention in our attempts to deal with this problem. Nominal filter capacity is 800 gpm each. Filter size is 13¼ ft × 25 ft, a 331¼ ft² rectangle. The filter bed consists of 9-in.

gravel, 3-in. torpedo sand, and 24-in. filter sand. We have changed un-
derdrains, removed the media, and then cleaned and checked under-
drains of other filters before adding new media, but the problem continues. Can Power readers give us any clues on what is causing our troubles? Is our problem a common one, with which we must learn to live? If
so, what are effective steps to take to keep sand from the mill?—RCS

Float the pump suction in clearwell

RCS's problem is an interesting one, and perhaps the source of the trouble is in
the location of the suction in the clearwell. RCS says nothing about the dimensions or stability of water level in the clearwell, but if the pump suction were
from the upper level in the clearwell rather than from the bottom of a sump,
the water going to the mill might be freer of sand. If the level in the clearwell is
not stable enough for a standard suction connection, it would be possible to
install one of several types of floating-pump suction connections which various
manufacturers can furnish. Should the source of the sand be the lake rather
than the filters, perhaps the same type of pump suction is advisable at the lake
intake.

H E Ferrill, *Alexandria, La*

Try a coarser sand in filters to prevent loss

The excessive amount of sand in the filtered water appears to be caused by loss
of sand from the filter bed, from RCS's statement of the problem. The filter-
sand size was not given, but presence of sand in filtered water can result from:
  ■ Too fine a sand, or a sand that has been recently crushed.
  ■ Cracking of the sand bed (the cracking may not be visible from outside
the filter).
  ■ Presence of mud centers in the sand bed.
  RCS could consider use of a filter sand that is one step coarser than the sand
now in service. A sand that is waterworn is better for the purpose than recently
crushed material. Cracking of the sand bed could result from improper initial
installation or too high a washing rate. Presence of mud centers in the bed is
the result of inadequate treatment before filtration, allowing excessively high
turbidity to be passed to the filtration plant. This is aside from the question of
whether the Accelator is undersized or not, which is another matter for investigation. Finally, treatment insufficiency could be due to inadequate
chemicals reaching the Accelator because of clogged feed lines or partly closed
chemical-feed valves.

G Ghosh, *Calcutta, India*

Is filtration rate too high, capacity too low?

First of all, RCS' Accelator capacity should not exceed 2.5 gpm/ft^2 area. If
coagulation makes use of alum, jar tests should be run to find the optimum pH

Filtration rates for two flows

Flow, Mgd	Gpm/ft^2 area per no. of filters		
	Four	Five	Six
4000	2.1	1.68	1.4
5000	2.6	2.1	1.75

and the best alum dosage. Continuous feed of liquid alum gives the accurate feed required when coagulating with alum. Unless coagulation is correct, floc carried over can occlude sand particles, difficult to remove from filters even at high backwash rates.

Also, filtration rates of over 2 gpm/ft^2 area tend to pull the sand through the sand bed, especially if the filtration rates are changed rapidly or maximum rates are exceeded and especially when a filter is out for backwashing. If RCS will set constant filter rates and watch his loss-of-head gages, there should be a decided improvement.

Insufficient first-stage Accelator capacity and inadequate filter capacity also appear indicated. For filters, see the table. At an 800-gpm flow, the rate would be 2.42 gpm/ft^2 area—too high. For the Accelator, a 42-ft-diameter model would have 2 gpm/ft^2 area at 5000 Mgd, and a 32-ft-diameter one would have 4 gpm/ft^2 area at the same daily rate. A suggested maximum rate of 2.5 gpm/ft^2 area would need two 30-ft-diameter units.

W B GURNEY, *Baton Rouge, La*

Piping setup to filters needs attention

To begin with, is the sand in question lake sand or filter sand? Analysis under a microscope might determine this. If RCS wants to find whether an appreciable amount of sand is coming in from the lake and whether the amount varies with conditions of weather, he might try a fine-mesh cloth "sock" suspended in the intake flume (see "a" in drawing.) A study of the piping to the filters might

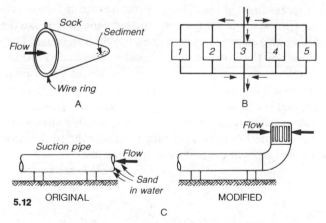

reveal a trouble source. At "b" (drawing) is a simple but particularly poor arrangement, in which filter No. 3 takes much of the flow because it has the least resistance in flow path to it. Filters 1, 2, 4, and 5 represent reserve capacity, which is, in effect, unused. It is altogether possible that the resulting excessive velocity of flow through filter No. 3 is eroding the filter bed, and that this is the source of the sand found.

To make the present system work, it will be necessary to equalize flows, so that each filter will carry one-fifth of the total flow. If there is no valving in the inlet to each filter bed, perhaps RCS could partly block inlets to 2, 3, and 4, especially that to 3, to obtain the right flow. If the sand is found to originate in the lake, RCS could try a different inlet, as at "c" in drawing. This would take water farther from the bottom, where it would be cleaner.

A M PALMER, *Brooklyn, NY*

Check on erosion vs corrosion

Perhaps RCS could enlarge his clearwell somewhat to give more settling time to clear the fine sand out of the filtered water. A large tank, if available, would serve the same purpose. I question that the sand did much damage to air-compressor cooling passages. Perhaps corrosion is the real cause of metal wastage in the passages. Water velocities in most compressor cooling systems are not high enough to scour out metal unless the abrasive particles are very numerous, large, or hard.

Erosion damage should also show up in general mill piping, at locations such as valve seats, pipe elbows, and opposite nozzle discharges. RCS could look there for erosion damage before making a final decision on the harm done. For individual machines and controls, cartridge-type filters are available to remove particles of the size of sand and silt. A small number of these filters will protect vital units at low cost.

D A RALIN, *Wilmington, Del*

Highest filter rate governs

The trouble appears to be in the filters, where the rate of flow may be too high at times. Lake Erie water is not generally turbid enough to make the sand in it a problem, although a poorly designed intake could be a cause. The maximum flow, and not merely the average flow, should be checked at one of the filters, before doing any other work. An excessive rate would be an indication that sand is being carried into the clearwell and is available for pickup by the mill pumps. If the sand is as fine as it usually is in these cases, additional filters in the piping system will not be a wise investment, because they will require too much attention. If another filter is needed to handle the peak water demands, RCS would be well advised to get it as soon as possible.

K F JACKSON, *Cleveland, Ohio*

6 Instrumentation and Control

6.1 A program for monitoring emissions

As the instrumentation supervisor for a large utility, I've recently been assigned the project of developing a continuous-emissions-monitoring (CEM) program for two large coal-fired plants. I'm not really familiar with most CEM instrumentation, but I've heard numerous horror stories about sample probes corroding and falling into the stack, sample lines plugging, and instrumentation failing because of temperature and vibration. I feel there's a good chance I might make the same mistakes as others. Where should I start to develop a dependable system? Will vendors help me choose the proper equipment for my stack conditions objectively, or should I talk to consultants? I'd like to know what potential pitfalls I should be aware of so I can avoid unnecessary cost or labor.

I've read that the most successful monitoring programs have been those with dedicated instrumentation technicians and management backing, but I've had trouble securing the proper support from my management. How should I go about convincing my superiors that I'll need more funding than our instrumentation group is getting to install and maintain a functioning CEM system? How much should I expect to spend for the installation, startup, and annual maintenance of a CEM system? What experience have Power readers had with designing and maintaining a continuous-emissions system? Is it possible to design a functioning system the first time? Are we dealing with reliable instrumentation backed up by dependable suppliers?—RPE

Recommends buying a complete package

Best starting point is familiarity with the EPA regulations or a talk with a consultant who knows them. Filing for an EPA permit may be necessary, too. After installation, an EPA compliance monitoring test and a compliance permit may be needed, as well. When talking to vendors, we tell them that we intend to buy a startup package and an EPA compliance test package. We talk to only those vendors which offer these packages. Vendors know the EPA regulations inside and out. We ask each vendor for a users' list of companies with a situation similar to ours.

Each of the two basic types of system has its own problems. Plugged sample lines occur on extractive systems, as against temperature, corrosion, and vibration problems for in-situ types. In addition, across-the-stack in-situ equipment has alignment problems. We bought two items helpful in maintenance:

■ A control-room display unit for analyzer output. Its alarm alerts us to minor problems that could turn into major ones if left unattended.

■ Vendor-school attendance for an instrument technician. The technician helps the vendor representative during startup and EPA compliance testing.

Our maintenance program consists of a daily check, by the instrument technician, of the control-room display unit. He also watches the system go through a calibration sequence. A well-trained technician can tell much from the response to a calibration sequence of no more than 15 minutes. Once a week, the technician and a helper inspect the instrumentation, doing routine maintenance and visual inspection, for an average time of one or two hours. During a lengthy downtime, a manual calibration check is performed. We keep a fair amount of spare parts on hand.

With today's EPA compliance standards, a reliable CEM system and trained personnel are a must. From management's standpoint, the system will not cut fuel cost or save money, but a surprise forced shutdown by EPA because of high excess emissions can be expensive.

R A GEDEON, *Shreveport, La*

We started at an annual APCA meeting

Answering RPE's questions one by one, I will begin with how we started out on our system. We first decided that we had to know what people were talking about. We went down to Miami in 1972 to the annual meeting of the Air Pollution Control Assn to spend an informative week. We met companies with firsthand knowledge and learned that all companies were just beginning to develop the instruments necessary for our monitoring. One consultant, recommending leasing instead of buying, suggested an instrument manufacturer, and we contracted for his equipment. We have had unforeseen problems, troubles, and expenses, but the manufacturer has cooperated and still services our equipment.

On vendors vs consultants for equipment choice, there are now experienced consultants who should be put under contract to study the problem. If RPE wants to start with instrument manufacturers, he should get lists of users and thoroughly investigate several installations. APCA annual meetings give a chance to meet consultants and equipment manufacturers. Last year, we found our new acquisition computer in this way. Avoiding unnecessary labor costs depends first on availability of manufacturer service people. RPE doesn't want service people who have to travel several hundred miles. A local service person eliminates travel and hotel charges. Of course, an unexperienced local man is as much a problem as one who is too far away.

Manufacturer training classes for your electronic technician are worth looking into. On another type of equipment, we traveled from Cleveland to Florida to discuss a problem with a manufacturer's service manager, but his primary help was an offer to send an experienced man to Cleveland for $500/day plus expenses. His secretary, however, told us that their dealers' servicemen spent a week's training session at the Florida plant and that she would be happy to schedule our electrician for a free week's course. We paid airfare, they paid all the rest. So our suggestion is—be sure to ask what might be available.

Incidentially, be sure that "installation charges" don't mean that the manufacturer will send someone to watch you do the work. We found out that a $1500 installation charge meant sending someone to watch us. The "engineer" had not watched before and didn't know the drawings well, because he had not installed a job before. A fixed-fee contract, with the manufacturer completely responsible for installation and startup, plus a 30-60-day operational guarantee, might be a good idea. Don't forget to consider a stack platform for any in-stack monitor. Although it may exceed the instrument in cost, it is essential for testing and service. Get approval of the platform from local, state, and/or federal EPA if possible. Remember, too, that a standard 360-deg platform is probably available for less than a custom-built 270-deg platform with just enough area.

If a high roof is available, select stairs to the platform instead of a ladder. Also, tests require lift of equipment to the platform, so provide for a lifting method. Spare-parts costs are another expense. We found that a failed lamp was reworked by the factory at approximately 50% of cost when given to the serviceman but only 25% when we shipped it to the factory on our purchase order. It came back faster, too.

Convincing superiors so that you can get more funds can take either of two lines. A report that considers best solution, cost quotation, and instrument literature is one way. An alternative instrument can be included, but if it can't do the job and is much cheaper, you may get stuck with an instrument you don't want. In the report, also detail service-contract cost and the technicians needed, and a summary of the problem-solving activities. The second choice, and perhaps the most practical today, is to contract with a consultant to study the problem, write a formal report, and make recommendations. Contract with a consultant who has done a similar study, and get a quotation for cost and time involved.

For costs of installing, startup, and annual maintenance, first step is to determine the pollutants to monitor. Do you want one instrument to do several jobs or a separate one for each pollutant? We have one instrument for five pollutants: opacity, SO_2, CO_2, CO, and NO. This has been satisfactory for us. Our 1980 cost with a one-year service contract was $35,000 for the instrument. Installation cost was about $5000. We had a platform and needed only to replace an older instrument. Probably $50,000 would be reasonable for an instrument today, plus installation.

Our instrument had a 5-in. strip-chart recorder, which we thought inadequate. Because we have 12-in. circular-chart recorders for all power-house instrumentation, we built a panel containing five 12-in. two-pen circular-chart recorders. On one recorder, we included total steam generated, since we were concerned that pollutant values would change with steam load. The panel cost about $10,000. Annual maintenance for the system is about $8000, including travel. Our equipment has been reliable, and we think we can depend on the manufacturer. We have run a CEM program for the past eight years, on a plant with two 100,000-lb/hr chain-grate boilers with fabric filter. Our new monitor records SO_2 (0-3000 ppm), CO_2 (0-20%), CO (0-500 ppm), NO (0-1000 ppm), and opacity (0-20%). It meets the requirement for zero and span checks which were not in the earlier model.

In addition to recorders on the firing floor, we transmit 0-1-V signals from recorders to office area, where a computer prints our data and plots daily curves. We have not experienced major corroding of our instrument, because the measurement bases on optical absorption spectroscopy instead of extracting gas samples that flow through the instrument. We have had problems with lamps and sensors for individual pollutants, but we have had no long-time outages.

We plan on at least one call a month to assure that the instrument is in calibration and free of problems. Our present computer system is a data-acquisition one with an integrating analog-to-digital system card for our stack monitor. The 1982 cost of our computer system was about $9000, including video terminal and printer. We use the computer for a power-demand monitor, too.

J W KIRCHNER, *Cleveland, Ohio*

Visit the cited references, ask questions

CEM is a tenacious problem at best. The field is loaded with equipment that does not work consistently, does not give reliable continuous readings, and breaks down constantly. A call to the manufacturer for information is a good idea, but there are other steps too. In deciding what equipment is best for an application, ask for references on applications in your type of situation. Then call, or better yet, visit the references. Ask them how the equipment is running, what problems they have had, and what the quality of service has been. How long did service take, and how often was it necessary? Support and spare parts from the equipment supplier is another question to ask. A visit to the manufacturer is advisable, too. Some manufacturers offer free factory training for users.

Traditionally, monitoring equipment does not give a "return on investment." The investment for basic equipment for opacity monitoring can be as low as $1500 per unit. Installation can run from $350/day and up. This is for field service alone, without expenses and travel. RPE should plan to purchase a system that has easy access for maintenance people, few moving parts, and a minimum of factory replacement parts, all of which can be easily handled by the plant staff.

M JEFFERYS, *Levittown, Penn*

Plan on an 18-month project by segments

Here is a plan for developing a CEM with an extraction-type monitor. Start with the first three months, in which the federal as well as state requirements concerning clean air are studied. These needs call for measuring NO_X, SO_2, opacity, O_2, etc. A good agency relationship is important because it helps highlight any unachievable standards, so that alternative solutions can be found.

■ **The first three months.** Find out the type and quality of the current fuels and what the existing flue-gas analysis is. Find the present management's plans to meet the CEM program needs, and whether mechanical equipment will be added or not. Find out what the fuel and flue-gas analysis will be about two years hence when you are trying to start up the CEM instruments. Design now for the worst quality of flue gas.

Conduct an industry and vendor survey on CEM systems in operation. Look at cost of instruments and extent of maintenance, and also reliability and extent of vendor response to existing installations. Look for problems with sampling and sample-conditioning equipment, and find out how widely the equipment has served for EPA reporting and plant control. The survey will automatically point towards the pitfalls in different systems—pitfalls that RPE should take into account in design of his system. The survey will also indicate extent of use of instrument vendors and A/Es in the CEM program.

Decide during this period on whether EPA compliance and reporting or EPA reporting plus plant control is to be the end use for the analysis. In the three-month period, at least two engineers will be necessary. The persons must understand fuel and flue-gas analysis and instrumentation. One engineer will collect in-house data and survey industry and the vendors, while the other becomes familiar with EPA and local ordinances, and coordinates compliance requirements with the agencies. A plant chemist or chemical engineer would be a big help here.

■ **The next three months.** This work should give an approximate idea of cost. Depending on constituents under analysis, the equipment alone will run $30,000 to $40,000 per constituent. Installation will run at least 50% of this. For engineering, estimate about $40/hr, no matter who does it. Cost of the computer/multiplexer depends on whether RPE has a computer. Startup cost, using the vendor, will run to about $500 per day, excluding transportation and hotel.

Now the budgeted sum can be decided on, along with the schedule, over 18 to 24 months. Plan manhours and manpower needs. Categorize separately engineering, procurement, installation, and startup. Keep a separate maintenance budget. Decide on whether the program will be achieved through in-house work, through the instrument vendor, or through an A/E. An A/E can do about 90% of the analysis and survey work in the first three months. He can also recommend a vendor, CEM sampling points, and the justification for both. Do the job of coordinating with local agencies and the analysis of fuel and flue gas in house, however. If the plant decides to do the engineering, get the services of an experienced computer engineer.

■ **Farther down the road.** During the next six-12-month period, specify the CEM instruments, computer, and multiplexer. Prepare instrument, electrical, structural, and mechanical drawings. Purchase the instruments, computer, and multiplexer. Review the drawings with construction contractors and award the contracts. During the 12th to 15th months, check and settle any interfacing problems with contractors. Plant-outage needs are an example of the problem. With construction complete, start up and test in the 15th to 18th month.

■ **To stay out of trouble.** First, keep the analyzer as close to the sampling point as possible. Next, remember that two sample-conditioning systems will make work easier, because of formation of acids and solids which are drawn into the sampling lines. These acids and solids are partly removed by filters in the stream, so keep enough spare elements handy.

Don't waste time trying to analyze the filters or sampling lines unless absolutely necessary. Replace filters frequently, perhaps daily or weekly. Replace sampling lines every six months to one year. If sampling at the breeching/ducts is acceptable, initial cost will be less and future maintenance will be easier. RPE can design a correctly functioning system the first time—provided he realizes that day-to-day maintenance is inescapable. Remember, too, that no instrument vendor will mention or emphasize maintenance.

RPE should also remember that the CEM analyzer system will be down every day for at least half an hour for automatic calibration. In addition, the analyzer may not read correctly or in range during fast load change, normally at evening every day. Downtime for filter change depends on tapping-point location but takes at least two hours. Some check pointers for analyzers:

■ Number of automatic or manual sampling points.

■ Distance of analyzer cabinets from sampling points.

■ Is sampling-point switching automatic? When switching occurs, does analyzer output go to zero or stay in last position?

■ Is analyzer calibration purging automatic? What happens to output then?

■ If analyzer output goes to zero in switching or purging, is there digital output or contact to blank the signal at analyzer or computer?

For an existing computer, you should check these:

■ Is there spare capacity to take analyzer signals?

■ How many signals are needed, analog and digital?

■ Signal strength level, such as 4-20 mA, dpdt, etc.

■ For computer calculations, are signals other than the analyzers needed? Number and type?

■ What action would be taken on the computer during calibration, filter changing, purging, or analyzer malfunction?

C BALASUBRAMANIAN, *Wallingford, Penn*

A knowledgeable consultant is a good shortcut, but stay involved in the project

We have just finished installing a continuous-emission-monitoring system and therefore may be able to help RPE. He could either have a new plant or be required by regulations to monitor and control emissions at an older plant. In either case, the schedule is important. The more time available, the less will a consultant be needed, even though a good one will be a help. A consultant helps by furnishing extra hours for the work of specifying and designing, without adding to the permanent payroll.

The consultant has been through the process before. He probably knows of some equipment suppliers who have had little success, and he can eliminate them right away. He will be familiar with applicable regulations and with regulatory agencies. He has written the specifications before and should be able to produce them faster than the customer can.

Even if a plant hires a consultant, however, the plant should stay involved in the process and not turn it over to the consultant. Be certain that you understand the regulations. Get the consultant to go through them with you, showing you how they apply to you and why other regulations don't apply. Investigate suppliers, too. All of them have had problems with some installation. Next come the references. Talk to engineering and purchasing people of the referenced firm. Ask if the supplier submitted shop drawings and descriptive material in time for orderly review by the regulatory agencies and for design of the installation. Ask if the contract documents were handled well.

Then talk to the reference's construction people about installation and start-up problems. Ask the operating and maintenance people how much time does the equipment require, how does it react to its environment, and did the certification work go well. Documenting the discussions will help show your management realistic estimates. We spent about $150,000 on a gas analyzer, an opacity monitor, and a data-acquisition system. Installation costs will vary with amount of equipment, where it goes, and what it needs, including access platforms, electricity, and perhaps compressed air. Our in-situ analyzer has a heated housing, but there is still a question of what happens when the housing is opened for maintenance or calibration.

For your first-year maintenance budget, allow about 5% of the purchase price for parts, even though you don't buy them all at once. Allow another 5% for manufacturer's service representatives' work. Most important is allowance for your own technician's time, at least 2 hr/day per set of instruments for the technician to check the equipment and hold its hand. Consider reducing the time only after a track record develops.

Give the technician good training. Don't expect him to learn everything by looking over the service rep's shoulder. The better the technician, the less will the service rep be needed. Vendor reps are paid for at $600-800 a day, including travel time, plus travel and living expenses. Remember that the rep comes when he finishes his current job; your man will be available much sooner.

Write detailed specifications. We were burned when a vendor misunderstood one of our requirements.

D TAYLOR, *White Bear Lake, Minn*

Extractive monitors preferred over in-situ

On our gas-fired boilers, we monitor opacity, CO/O_2, and SO_2. The opacity instrument is an in-situ one. The other two are of extractive type. We regularly clean the lens and calibrate the opacity monitor. Zero check is a problem, however. For the extractive instruments, we change filters regularly. Primarily because of our preventive maintenance and our relatively clean fuel, we have not had major problems with our CEM instrumentation. RPE should consider the following:

■ Management must support a CEM program if the program is to be effective. The company environmental department can help justify the costs of added maintenance. The company should try hard to develop a cooperative relationship with regulatory agencies, too.

■ We have preferred extractive monitors. In coal service, the conditioning system needs careful design. Sample lines should be ⅜ in. to prevent pluggage. Easy access, isolation, and cleaning are important.

■ Some conductivity-based (such as SO_2) and IR analyzers require stable temperature environment for proper functioning. The stability also increases life of electronic components. We have learned that some utilities keep their CEM instrumentation on dedicated power supply for high reliability.

■ In-situ monitors are more likely to be affected by vibration and heating and cooling of the stack. And misalignment problems often result from vibration, so care is needed in installation.

G C SHAH, *Houston, Tex*

6.2 Improve piping on h-p instrument lines

Instrument piping seems to be one of the minor areas in piping that doesn't get the attention it deserves. We have been looking into ways to upgrade our instrument and control piping, starting with the simple and obvious improvements in routing and valve and fitting selection. Along the way, we have come across several problems that no one seems to know much about. The question of blowing down instrument lines is one such topic. We have tended not to add valves for blowing out or purging lines that are not subject to flow, but we have read recently of opinions that regular cleanout is desirable.

Also, temperature of instrument valves and lines is not clear-cut. Should a plant figure on full steam temperature on lines, fittings, and valves, or only the actual temperature? Sizing of line and valves is another question. Should we specify a safe minimum or call for small-bore stuff if theory says it will work? We will appreciate hearing from POWER readers on these and similar I&C matters.—GRA

Blowdown valve guarantees clearance

An instrument line should have a blowdown valve for cleaning when the line becomes fouled by oil or other contaminants. Also, a technician sometimes wants to assure himself that the line is clear. If the instruments are not compensated, then the line must be at full temperature for short runs to the instruments.

No line should have a rating less than the temperature or pressure of expected operation. Safe minimum for line sizing is the manufacturer's recommendation for his product.

T W HAMMOND, *Santa Nella, Calif*

General specifications are helpful

Feedback from several operational plants in the past decade has spurred us to set up general specs for the instrument impulse piping system. The specs largely answer GRA's questions (also see sketch, 6.2A):

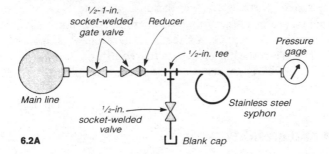

■ Blowing down the line from the stagnant part of impulse piping is strongly recommended. It is the only way to keep small-diameter instrument piping from choking on corrosion product and mill scale. These troublemakers readily form, especially during early operation.

■ Provide instrument impulse piping in only three sizes: ½ in., ¾ and 1 in. NPS. Pick the schedule—whether 40, 80, or 160—as required by the process. Although ⅛ in. NPS has been in many installations, plugging troubles have led to its removal from our specifications.

■ The economy of providing different grades of material in different sections of impulse piping is questionable. Material of all piping, fittings, valves, etc, on impulse routing should conform to main process parameters. This holds for uninsulated loops, too, which are normally full of condensate and at lower temperature. This is a typical setup, illustrating the general principles outlined above.

B K MITRA, *Philadelphia, Penn*

Blowdown is accepted practice

Regular cleanout is certainly desirable—probably the most common malfunctions in instrument piping are caused by dirty lines. Blowdown is the accepted practice, via such arrangements as shown in sketches 6.2B and 6.2C. If cleaning agents are used, be sure the pipe material and size are taken into account.

F R Murphy, *San Diego, Calif*

Put heavy half-couplings on main line

For high-temperature and pressure locations, main-pipe connections for pressure indicators and thermowells should be 3000-6000-lb half-couplings. Do not put ordinary pipe couplings on this service. In our plant, we had to switch to quarter-turn ball valves at several locations after we had had frequent leaks with needle valves. These valves were on sampling service. Ironically, the leaks' cause was excessive stress on valve plug and seat by operators who wanted tightly shut valves to prevent leakage.

Bleed valves on instrument lines are valuable for blowing down and cleaning during startups. Vents to remove trapped air at high points are also necessary. Most of our I&C tubing is ⅜ in. or ¼-in. stainless steel. Tubing size is usually governed by size of porthole connections on transmitters, I/Ps, actuators, etc. We nevertheless prefer ⅜-in. tubing, with length kept short, where freezing or pluggage is possible. The quick-connect fittings often found in bomb-sampling work need pressure-relieving means, actuated after a sample is taken. High pressure under quick-connects not only makes sampling difficult but also damages fittings.

G C Shah, *Houston, Tex*

Support and stability are keys

Although most gages have ¼-in. stems, this does not mean that the entire piping system should be that size. My own choice is for ¾ in. at the instrument location. This size gives support and stability and is easily blown down.

C R Klein, *Wauwatosa, Wis*

Standardize root connections and valves

Here are some ideas that GRA may find helpful:

■ Standardize root connections and root valves on 1-in. nominal pipe size. Make all root valves gate valves, so that lines can be blown down, drained, filled, and hydro-tested from the root connection.

■ Select valves of the same or higher pressure/temperature rating than for the line that they tap. On high-pressure steam lines, make the root valve a Y-pattern forged steel body valve. Two valves in series will give added assurance.

■ Connection fittings on the line should be the same material as the line tapped. Keep the number of fittings in root-valve lines to the minimum, because socket-weld or screwed fittings provide pockets for dirt.

■ Specify at least Sch-80 pipe on root-valve lines, and be sure to ream the ends before making joints. This prevents dirt from becoming pocketed behind a burr at the cut end of a nipple.

■ Look into the new "one-piece" root valves, offered by at least two manufacturers. The valves have a nipple and weld boss forged integral with the body.

■ Put gage taps at the top of the line if possible, to prevent the root connection from becoming a drip and dirt leg. And set a screwed adapter between the root valve and gage line (sketch, 6.2B), instead of a socket-weld adapter. This allows blowdown of the root valve through its full opening instead of through a piece of tubing.

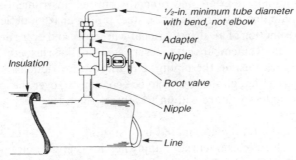

6.2B

■ Smooth bends are better than elbows, because the drilled passages of elbows, and of tubing fittings as well, can let dirt accumulate at low points. Support tubing properly. Sagging in low-flow or static lines can let dirt build at low points. For tubing, put in enough straight connectors so that the tubing can be disassembled for periodic blowdowns.

■ At the instrument connection itself, put in two valves—an instrument valve and a blowdown/test valve (sketch, 6.2C). The second or blowdown/test valve allows the instrument line to be blown down without disturbing the instrument. In addition, especially at pressure-transducer connections, a test gage can be set on this valve.

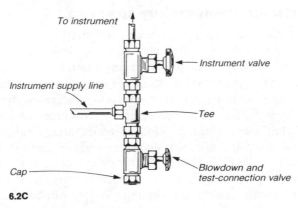

6.2C

■ Make up a blowdown line for needs of several locations. An "idiot" feature—a cap on the outlet of the blowdown valve—is a must.

For valves, specify 300 WSP screwed bronze, union bonnet, regrinding seat, or wedge gate, for saturated-steam pressure to 200 psig. Above 200 psig or for superheated steam, call for forged steel body gate valves, Stellite faced, OS&Y, and bolted bonnet.

J M MICHAELS, *Brooklyn, NY*

Compressed-air lines need attention, too

Although not really high pressure, instrument-air systems have problems that may be troublesome to GRA and others. Common instrument-system (air) problems include line rust from carbon steel piping or from deficient galvanized pipe. Malfunction of air driers can put water, oil, and even silica-gel particles in the system. Remember that on every startup, line rust, etc, is driven out to the remote locations on the piping, building up restrictions. Never install Sch-40 fittings. Although they seem to be economical, over the long run they really are not, unless you are willing to subject them to a very good quality-assurance effort.

Line rust can be totally eliminated by installing Type 304 or 316 stainless pipe. A fully welded instrument-air supply header is preferable, although taps can be by screwed fittings. Run 2-in. or 1-in. air headers where possible. Never put in a needle valve, and never specify a ¼-in. valve, of any pressure rating, for instrument-air lines. A ½-in. valve is necessary.

Instruments over 500 ft from the compressor should get an additional drip well in the line. Do not pipe more than five field instruments from a single ½-in. tapoff, and if possible reduce pressure for all at one stage. If the instrument-air compressor is on a floor below several floors of instrument locations, air-header zones will be at various temperatures simultaneously, so that an intermediate header with drain will be needed.

C BALASUBRAMANAN, *Philadelphia, Penn*

Apply maximum-temperature spec to all

Instrument piping has undoubtedly been overlooked in many plants. Often, sizing and routing are left to the contractor to settle in the field. Future maintenance needs get no thought, and many control problems are traceable to badly sized and installed piping. All steam instruments need blowdown piping, but the lines must be placed properly to prevent overheating the instrument (sketch, 6.2D). Flow and pressure transmitters need a valve at the device to isolate the instrument during high-temperature steam flow. Periodic blowdown of impulse lines eliminates plugging and replacement of piping.

For size of impulse piping, ½ in. O.D. tubing—or equivalent pipe size— should be good for most applications. Remember that if even a small leak should occur near the instrument, the impulse line will be at close to steam

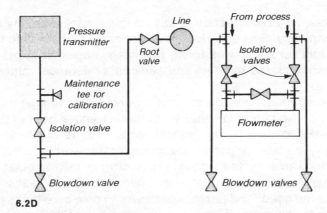

6.2D

temperature, so apply maximum-temperature specs to all lines, fittings, and valves, for safety. Before deciding on impulse-pipe routing, review these experience-based suggestions:

■ Slope steam impulse line about 1 in./ft toward the instrument to prevent air pockets.

■ The installation should allow lines to expand during blowdown.

■ Select line location to protect from freezing winter drafts. Remember that there is no flow in the line.

R W Spellman, *Commack, NY*

Small-bore tubing has good record

Blowdown should be a must before final connection at any element of an instrument system. Regular in-service blowdown, however, should be examined in light of possible benefits. If experience or logical reasoning indicates deposits, buildup, or trash accumulation in the lines, then set up a regular blowdown and/or line-cleaning procedure. These problems are often the result of poor tubing-material selection. An example: Draft-gage manometer tubing may be exposed to SO_2, H_2SO_3, and weak H_2SO_4 from the flue gas. The one stainless steel that will resist all three of these corrosive agents is Alloy 20. The 300 and 400 series are poor choices for the weaker acids.

Line sizing is also a matter of choice. "Small-bore" tubing has had considerable success. Flow in these lines is minimal, so that engineering concerns for pressure drop and head loss are nonexistent. In systems where the readout or controller is some distance from the detection location, my preference is for P/A or P/V transducers, with standard electrical interconnections.

B B Brown, *Hagerstown, Md*

6.3 Air or electronic instrumentation?

In my initiation to steam-plant engineering (lasting three years so far), I have had little time to wonder about the overall picture. Project details, breakdowns, new construction, and testing have taken up my attention nearly completely. One question that I first came across some time ago, however, has set me to wondering how many engineering policies and practices are dictated by whim and personal preference, rather than by real analysis.

In our operations, we are measuring pressures, temperatures, and flow rates, and controlling these and other variables both in the steam plant and out in the process departments. I found out early that the steam plant stays largely with pneumatic instrumentation, but that the process departments are largely on electronic instrumentation. This means that repair and maintenance work is hampered somewhat, because of the need for different specialists to take care of the results of differences in philosophies.

Why this condition should exist has been explained to me by proponents of both schools, but I can't see why either air or electronic cannot do for the whole plant. What are the thoughts of Power readers on this? Can the entire job be handled by one type of instrumentation? What is the trend in this?—WS

Lack of systematic, quantitative data is factor

Our boilers are pneumatically controlled, and our process side has electronic controls. It is true that the choice is often the result of personal preferences rather than real analysis. Lack of systematic, quantitative data could be one reason for apparent reluctance to go through a cost comparison on these two methods for control. Qualitatively, here are some of the arguments:

■ **Cost.** On the average, electronic instruments are roughly 25-30% more expensive than equivalent pneumatics. Installation costs depend on distances, control-room requirements, and the presence or absence of a computer. Longer distances tend to increase tubing installation cost; cable trays for electronic systems make for a cheaper installation. As for computers, they interface easily with electronics by A/D converters. Pneumatics, on the other hand, require P/I or P/V converters in addition to A/D.

■ **Safety.** Pneumatics are naturally safer than electronic instruments, and do not suffer from electromagnetic interference (EMI) that is troublesome to electronics.

■ **Time lags** (dead time). Pneumatics are slower than electronics. For short distances (under about 500 ft), this difference in response speed may not be important. Over longer distances, long dead times can make a control highly unstable.

■ **Sensitivity and reproducibility** (accuracy). The two approaches are about equal for all industrial applications, in my opinion.

■ **Maintenance.** At our plant, instrument technicians are well experienced in both electronic and pneumatic instruments. "Replace-the-card" philosophy greatly simplifies electronic maintenance and does not call for extensive training. Technicians and union agreements have to be considered at each plant, naturally.

■ **Reliability.** In pneumatic systems, this depends directly on the reliability of the air compressor. In electronic systems, battery backup gives reliability.

There are other factors, too, such as complexity of the control schemes, vibration, and present inability of several variables, such as composition and pH, to be measured by pneumatics.

G C SHAH, *Houston, Tex*

Looks like both types will be around

WS's observations concerning standardizing on one type of instrumentation are admirable, especially from a maintenance and spares standpoint. There are times, however, when one type of instrumentation gets preference over the other because of the area and service involved. Powerplant reliability is a key issue here.

In many cases, the boilerhouse environment is much too harsh for electronic instrumentation. High temperatures, vibrations, dust, and so on, can play havoc with sensitive electronic gear. Pneumatic instrumentation has proven itself in powerplant service over the years, and it will consequently be around for some time to come.

In the realization that fuel supplies are finite (and fuel prices are not), efforts are now being made to improve combustion control and boiler efficiency. Microprocessors help in improving accuracy and efficiency of boiler control. This is an area where electronics can play a big part. It can be said, therefore, that the trend is not toward displacement of one type of instrumentation by another, but toward a convenient marriage of the two types.

R S DUBOVSKY, *Disputanta, Va*

Thinks slow trend is toward electronics

Direct answer to WS's question is "Yes, a single process plant can standardize on either pneumatic or electronic instrumentation for the whole plant." Economics, however, has dictated the selection of pneumatics in some areas, electronics in others. Both electronic and pneumatic control systems have pneumatic-actuated control valves. Therefore, at least one extra device (an L/P transducer) is necessary when the controller is electronic. Control loops that have little manual operator intervention and setpoint changes are economically handled by a local pneumatic controller/sensor system. Many such control loops are in the powerhouse (for example, DA level control, feedpump recirculation, steam-pressure reducing service).

Electronic controls are superior in performance to pneumatics in systems where operators can control in the manual mode and adjust setpoints from a central control-room location. In these applications, electronic controls take less power, respond faster, are more reliable, and adapt better to complex control strategies. Many of the control loops in the process-plant area fall in this category. The trend is toward complete plant-wide control with modern electronic distributed-control instrumentation. Nevertheless, pneumatics will be difficult to displace from the applications where it performs best.

D FLETCHER, *Charlotte, NC*

Pneumatic is more reliable in adversity

In the boilerhouse, we can't do the job without air power. That is simply the fact, no matter what modern electronic marvels are doing out in the plant—or out in space, for that matter. There are so many applications where we need heavy push or pull, fast action, and good resistance to environment. And, believe me, the general situation is not getting better so far as any of these requirements go.

Take the boiler backend, for just one example. With bigger units, more gas flow, more heat recovery, and all kinds of pollution-control equipment, the demands on control and actuation are going up steadily. We have to put in valves and actuators that stroke full in a few seconds and that can resist all kinds of weather and all the corrosive vapors and liquids in the space between boiler and stack.

If we installed electrical controls and equipment here, we would have a never-ending problem with wiring, panels, switch boxes, and motors. We have the problem right now of having to put in pump drives around coal piles, ash pits, scrubbers, and so forth. The smaller we keep our exposure in electronics, the better the reliability.

We keep our control air clean and dry, of course. This is merely a small dues payment to join the club, as we look at it. Special monitoring and high-level control, done inside the control room and at well-protected machinery, is different, naturally. There you need electronic instrumentation—the more the better, evidently.

H E HENDERSON, *Baltimore, Md*

Powerplant engineers themselves need updating

In 40 years' experience with powerplant operation, I have found that troubles with instruments start when the designers of the plant lack complete knowledge of the instruments needed. Later, engineering personnel have to evaluate the need for instruments or make any changes required. This calls for a close relationship between management and the operating staff.

Calibration is important for all instruments. This function and others are described fully in the manuals, but, unfortunately, the manuals are often given to plant operators and lost, so that it is necessary to get more manuals from the manufacturer of the instruments. There is a tendency among powerplant engineers to depend on local repair shops for instrument work. Often, the cost of operating in this manner is higher than if the plant's engineers were to take more interest in the instrumentation. Visits to other plants and exchange of information could help in this area.

L W THIELE, *Buffalo, NY*

6.4 Instrumentation vs water treatment

Here is a problem that perhaps will never happen—if we take the right steps now. Focal point is our water-treatment plant. We pretreat our water with a clarifier-softener and filtration unit, followed by demineralization with cation-anion trains and a mixed bed. The whole setup looks very complete, with recording charts, automatic cycle control, conductivity, pH sampling, etc. Apparently, the system designers took almost everything into account. Nevertheless, my experience with old-style equipment, where an operator monitors exhaustion, filter-bed condition, and ion-exchange-resin performance, and operates the valves himself, makes me fear that if something goes wrong with even a minor component in our system, we could be in for big trouble.

In my short study of the plant, I could see several applications for alarm instrumentation—on pH, conductivity, and oxygen, for example. We don't want to go ahead and spend money unnecessarily, however. I am therefore asking POWER readers: What has been your experience on highly automated water-treatment plants? Does lab analysis provide results quickly enough to allow for fast corrective action? Where would I expect to find problems in the resin-regeneration procedure? What added instrumentation have you found advisable? And can we get cost-effectiveness and higher reliability from add-on equipment?—SEY

Lab analysis backs up instruments effectively

SEY's apprehensions are justified. If something goes wrong with even a minor component in the typical state-of-the-art facility, big trouble is possible. Our demineralizers (two mixed-bed units and one cation/anion train followed by a mixed-bed polisher) have given many years of reliable service. Alarm and shutdown capability is a must, however.

Lab analysis of pH or conductivity can give results fast enough for corrective action on problems, but silica analysis takes more time. The general key to good operation is periodic lab analysis as a backup to on-line instrumentation for calibration. You have to be able to believe what the instruments are showing, and the electronics for alarm or shutdown must react fast enough to measured concentrations. Among the common problems in resin regeneration:

■ Regenerant dilution water flow.
■ Regenerant eduction timers.
■ Multiport-valve leakthrough.
■ Pump delivery of wrong amount of acid or caustic, even after several

strokes. This may be the result of failure to vent settling pots on suction side of pumps back to storage tanks.

■ Loss of resin, especially in backwash if screens are defective or if excessive amount of fines is lost.

■ Too low a caustic temperature.

■ Regenerant channeling in the vessel, usually the result of poor distribution-manifold design.

Cost effectiveness from add-on redundancy depends on the process. If the effluent stream goes to an intermediate storage tank for monitoring, you have about the best protection available. Cost/benefit depends on the damage that can be done at the process end. Is it strictly a loss of water? Will off-spec product result? Will contamination affect multiple units? What will be the result of contamination—in loss of generation, boiler-tube damage, cost of replacement power?

SEY could also determine his silica breakpoint and set his capacity at several thousand gallons less. (Silica analyzers, despite claims, are notorious maintenance items.) Also, periodically verify conductivity and pH readings. And remember that most mistakes happen in the manual mode for regeneration, so think everything over three times in that case.

J P Vergura, *Loudonville, NY*

Record-keeping is important for top performance

Highly automated water-treatment plants can be operated with top reliability. They still require some operator monitoring, however, especially during the regeneration cycle. Data-form record-keeping is most important. It starts with chemical feed rates to the clarifiers, and it extends to frequency of filter backwash. Monitor the demineralizer itself for backwash rates, chemical injection rates and concentrations, and rinse requirements for each bed. Backwash rates will fluctuate with water temperature, but proper instrumentation coupled with lab analysis should give fast enough corrective action. Recording of conductivity trends will indicate breakthrough on the anion and mixed beds before exhaustion, allowing correction.

For backwash, monitor rates for good bed expansion and removal of resin fines. Adjust valves as temperature varies. Proper concentrations and displacements of regenerant chemical going to both cation and anion beds are important, too. Concerning rinse cycling, which assures proper rinsing of beds after chemical injection, the outlet headers from all three beds should be located to ease backwash sampling and monitoring, activities that determine proper efficiency and fines removal.

This is what I consider the minimum for instrumentation and sampling in an automated water-treatment system:

Cation:

■ Influent meter or totalizer

■ Sampling from dilute chemical injection line

- Effluent sample line
- Adequate sample point at backwash outlet
- Flowmeters for backwash, chemical injection, and rinse rates

Anion:

- Sampling from dilute chemical injection line
- Effluent sample line
- Temperature monitor on chemical injection line
- Adequate sample point at backwash outlet
- Flowmeters for backwash, chemical injection, and rinse rates
- Conductivity monitoring of effluent, and in-bed monitoring to detect conductivity break during service run

Mixed bed:

- Influent meter to determine service run, and flowmeters for other rates
- Effluent sample line
- Sample lines from dilute chemical lines
- Temperature monitor on anion regeneration line
- Adequate sample point at backwash outlet
- Monitoring of conductivity and pH of effluent
- Metering alarm for end of service run

D GUENTHNER, *Hayden, Colo*

'It rained rootbeer' after this incident

Perhaps an incident we went through last year will show what can happen when instrumentation is incomplete. We treat water with a 500-gpm clarifier-softener, followed by two gravity filters, two carbon filters, two cation-anion trains (200 gpm each), and one 400-gpm mixed bed.

Early one morning, during the midnight-8 a.m. shift, operators had finished regenerating train A of the water plant. Both condensate-tank levels were high, and a boiler unit was at reduced load during condenser waterside cleaning. There were only about nine counts left on the mixed-bed flow accumulator clock, so the two operators decided to begin regenerating the mixed bed right after finishing train A.

With train B left in service and train A valved out, we started the mixed-bed regenerator cycle. An operator saw good resin movement inside. Near the end of the three-minute "settle" step, an operator opened acid and caustic regenerant sample valves. When the "acid in/caustic in" step began, he sampled both streams and adjusted flow rates until Baumé value was correct for both regenerants. Because we had had trouble with this step a few nights before, the operator repeated the regeneration and held the "acid in/caustic in" step for an extra 10 minutes. He found that the mixed-bed conductivity rinsed to low values very quickly at the end of the regeneration cycle.

After the extra 10 minutes, the operator switched from "hold" to "automatic." A Baumé check right then still looked good. The next step had been under way for about four minutes when the operators saw a foamy brown liquid pouring out of the mixed-bed drain pot and going into the drain trench. One operator could feel heat from the corner where the mixed-bed tank is. The foam was hot, with a sharp irritating odor, not acidic but like something burnt. Neither the water-plant operators nor the turbine-electric operator had ever seen anything like this in the water plant.

Opening the acid-regenerant sample valve gave a syrupy brown liquid for a few seconds. Then the cleared sample gave Baumé 6.2. Opening the caustic-regenerant sample valve gave a liquid too hot to measure Baumé. It sputtered and spat when it splashed into the drain trench. Acid and caustic pumps were shut down, but the operator knew that he couldn't leave the resin bed in that state, so he tried to cool the tank by manually indexing the regeneration step to "slow refill." Then things went wild. The tank contents seemed to boil. Punching the cycle to "fast refill" only made the tank hotter, bubbling out foam and slop. In a few seconds, the operator punched through to "final rinse."

Things really blew up then. Foam was coming out of the pot so fast that it rebounded straight up and "rained rootbeer." Two operators were splashed with acid and had to shower. The conductivity sample line, oscillating back and forth, sprayed foam around the water plant. Water from the line entered an electrical box near the silica analyzer, with a flash and sparks. The line broke, but an operator was able to shut off the valve.

The operators punched the cycle to "standby," opened a valve to drain the line to the condensate tanks, and left the water plant to get help. When they returned, there was a brown foamy material in the drain trench and an acid odor in the air. The mixed bed was still kicking out crud, but more slowly. A little while later, we found that the resin had been blown from the mixed bed. The 50-mesh, 316-stainless resin screens covering the 18 underdrain laterals had failed, too, allowing resin to flow out the rinse outlet valve. Resin in the bed costs about $7000; we had to replace it all after this incident.

Then we started to look for the cause of the trouble. Water-plant flows are recorded on two circular charts, which became keys in our search. One chart, for train A, monitors raw-water, train-A, and mixed-bed flows. The other chart monitors train-B, acid-dilution-water, and caustic-dilution-water flows. Pens are offset, so we had a little difficulty in determining absolute time. What we finally came up with indicates that:

■ Dilution-water flow was lost during the "acid in/caustic in" step, but the acid and caustic pumps continued to run.

■ Concentrated sulfuric acid and concentrated caustic soda were pumped into the resin bed for perhaps 13 minutes.

Assuming that about nine gallons of sulfuric acid and 10 gallons of caustic went in, we figured from a handbook that diluting 66-Baumé acid at atmospheric pressure would give a maximum temperature of around 268F. Diluting 50% caustic would give a temperature as high as 175F. The maximum rise from neutralizing the sulfuric with caustic would be approximately 285 deg F.

When the concentrated chemicals reached the resin bed, dilution with water in the tank caused extreme heating, which probably broke down the resins and started the foaming and odor. Indexing through to slow and then fast refill probably didn't make much difference, because there was no flow from train B except for a few flow spikes noted on the chart.

We faced some questions that needed answers: (1) Why did dilution-water flow stop? (2) Why didn't the automatic valves close to stop the concentrated acid and caustic flows when the dilution water stopped? And (3) Why was there no alarm to warn operators that the dilution flow had stopped?

In answer to (2) and (3), I can say that the conductivity probes that monitor acid and caustic concentration have been disconnected for a long time. These monitors were designed to signal incorrect conductivity, and possibly to stop acid and caustic pumps and close block valves.

The reason for removing the conductivity monitors was that they wasted acid and caustic regenerant, discharging as they do into the trench. Incorrect readings had been another reason. Hooking up the monitors again with the resulting loss of a little acid or caustic is a small price to pay for reliability.

Question (1) is still unanswered. Since one of the carbon filters went into automatic backwash and rinse at about the same time that dilution water was lost, perhaps the backwash flow lowered the clearwell level below the pump-suction inlets. One man here thinks that flow spikes seen on the charts during the trouble period were the result of the pumps cavitating at low suction heads. The alarm for low clearwell level that would have alerted operators to this had also been disconnected.

The lesson for SEY is that you can't neglect instruments, disconnect protective devices and alarms, or improve too much on the water-treatment companies' designs. If you do, your operators can be caught in the middle of some real messes, with no warning that they are headed for trouble.

T FITZSIMMONS, *Bismarck, ND*

Install instruments properly to forestall problems

A turbidimeter with high/low alarms on raw-water clarifier effluent gives a good indication of sludge removal. In normal operation, a flow totalizer initiates the sludge-removal cycle. In addition, the rake drive should include high-load interlocks (shutting off the motor), and low-load and no-load alarms. We find a conductivity analyzer, with high alarm and backup by lab analyses for silica, sodium, and pH, to be adequate. The regeneration cycle is started manually, based on either total gallons demineralized or conductivity alarm. We have traced many of our regeneration problems to leaky caustic and sulfuric valves. Stay away from butterfly valves. Another trouble source has been too-cold caustic solution.

SEY should install instruments properly, too. Flow glasses to get a flow check are a simple way to prevent many problems on flow-through instruments, such as pH meters, conductivity meters, or turbidimeters. Mechanical deficiencies cannot be cured by instruments. Remember that a thorough inspection of internals—distributors, support plate, baffle, and resin charge, among others—will eliminate many a surprise problem during startup.

G C SHAH, *Houston, Tex*

Experience says go with manually operated valves

In my experience on highly automated ion-exchange units, troubleshooting is often a problem. Resin analysis in the lab takes too much time to be useful at the time of trouble, for instance. If there is a problem in regeneration, the effect is felt only when the improperly regenerated resin passes through the loading chamber. By that time, much resin has moved. Malfunction of valves would give improper regeneration.

My advice would be to avoid excessive instrumentation in ion-exchange units and to stick to old-fashioned, manually operated valves. Some alarm instrumentation is needed, however. Examples would be a sodium analyzer on cation effluent, and a chloride analyzer on anion effluent. Conductivity and pH should be monitored continuously.

P TAGORE, *Brandon, Man, Canada*

Internal inspection is always a good idea

Install visual and audible alarms, local and remote, for all facilities. This will give quick identification of any malfunction that would otherwise not be known until lab results came in. Because lab results are usually intermittent, continuous monitoring with instrumentation gives higher reliability. Centralized panel-mounted annunciators indicate high/low levels in storage and chemical-solution vessels. Alarm sensors for end-of-service run of filters and ion exchangers, for high and low pH, and for conductivity are recommended.

Change in various flow rates, and in grade of regenerating chemical, are other possibilities for problems. And internal causes can occur, too. Internal inspection should include detection of slime and material on bed top, uneven bed, and performance of regenerator distributor. Send water through the inlet distributor to make sure that the spray pattern is symmetrical. If not, passages are clogged.

H B WAYNE, *Jamaica, NY*

6.5 How much stack instrumentation?

Four oil-fired boilers, ranging in size from 80,000 to 250,000 lb/hr. deliver 250-psig steam to our process operations. Our newest boiler has an efficient combustion control and carries a nearly constant load. Our load varies seasonally by a factor of 2 to 1, but the day-to-day load is steady. Our other boilers lack really effective combustion controls, but we are going to install this equipment.

Studies of the proposed installation have caused me to ask whether we could benefit from stack instrumentation, in addition to the basic combustion controls. Another consideration here is the possibility that we will be subject to strict regulation of air pollution. Up to now, our stack instrumentation has been only a temperature gage on one stack. Our management has not been enthusiastic about boiler-house instrumentation in the past. I need practical advice and experience on this, to convince a front office that has given instrument salesmen a hard time over the years.

Can Power readers tell me what types of instruments would be most advisable for our stacks, to help us with efficiency improvement? Where in the stack should the instrumentation be put? Can we cluster the instruments, or do we have to locate some near the breeching and others farther up the stack? Should we install SO_x and NO_x equipment? Will we need special training, tools to take care of the instrumentation?—MT

SO_x meter is not always needed, depending on fuel

If MT fires his boilers with only one grade or quality of fuel, the SO_x in the stack gases will remain in constant proportion to the fuel sulfur content and the firing rate. SO_x can then be calculated from analysis of the fuel and combustion rate. If local pollution-control regulations require lower SO_x discharge during certain atmospheric conditions and if all or part of the boiler load is carried by burning a lower-SO_x fuel to meet the regulations, then an SO_x meter can become advisable.

A smoke-density meter is a must. Give preference to location at a point of negative draft to reduce the tendency of flyash to settle in the sighting tube and give a false high reading. The meter should record on the O_2 and combustibles analyzer chart.

B M Kine, *Vancouver, BC, Canada*

Simple controls boost efficiency with O_2-trimming

The most efficient boilers operate at the lowest possible excess oxygen. Generally, industrial-boiler manufacturers guarantee their products at 3% oxygen. We have been able to improve on this by O_2-trimmed combustion controls, proper tuning of these controls, and good operating practices. This has resulted in operation of our boilers at between 1.2-1.8% oxygen. Resulting saving in fuel alone is considerable.

Sketch (6.5A) shows a simple combustion system with O_2 trim, working as a parallel positioning system in which the master density adjusts the forced-draft control drive and also the fuel-oil valve. The fuel/air ratio system adjusts to trim the fuel-oil flow. The oxygen analyzer output adjusts the air-flow signal into the fuel-control loop to maintain the oxygen set point.

O_2 measurement alone is not enough. Combustibles and smoke (opacity) should be measured. Reading of combustibles alerts operators to improper burner performance. The smoke (opacity) meter allows boiler operation at a low-excess-oxygen level just below the point at which the boiler begins to smoke and exceed local EPA limits on opacity. The oxygen and combustibles analyzer should be sampling the boiler exit gas upstream of the air preheater (see sketch, 6.5B). Downstream of the air preheater should be a sampling point to measure oxygen with an orsat. This allows air-heater leakage to be determined. To find boiler efficiency (by ASME short form), MT needs temperatures of the air entering and of gas leaving the boiler and air preheater.

SO_x should not be a problem unless the fuel is a high-sulfur variety. NO_x may be a problem, however. One of the many stack-emission consulting firms could help here, measuring stack emissions and comparing them with state standards. Continuous monitoring is normally not necessary. A suitable location, ordinarily one-third the way up the stack, is all that may be needed. And if MT runs into an NO_x problem, the boiler manufacturer will be able to help correct deficiencies.

D F Madura, *Mystic, Conn*

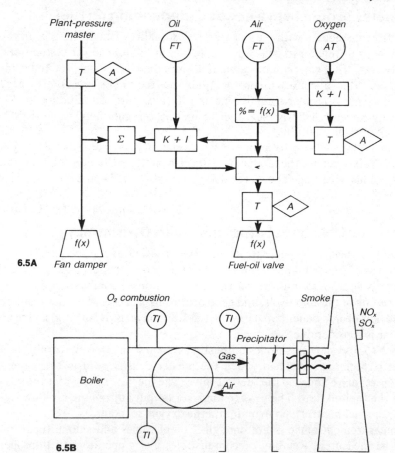

6.5A Fan damper   Fuel-oil valve

6.5B

Aim for operator skill, know-how, motivation

The two most useful aids for efficiency improvement are an O_2 monitor and a device for measuring flue-gas temperature. To reduce sampling-line-maintenance cost, get the type of O_2 monitor that mounts in the flue-gas stream. A gas-filled coiled-tube type of temperature-measuring device should be adequate. Both O_2 and exit-gas temperature should be recorded continuously. A stack-opacity meter could reduce the possibility of receiving an air-pollution citation. I would not recommend NO_x or SO_x instrumentation. Most areas do not require it, and it is useless for efficiency improvement.

The O_2 monitor and the temperature-measuring device can be placed together in the boiler-gas outlet duct or in the economizer outlet duct if the units have economizers. If the units have regenerative air heaters, place the O_2 analyzer after the boiler bank, and the temperature device after the air heater. Install a series of taps in the same general area for calibrating instruments against actual duct conditions. Set taps to allow measurement of flue-gas temperature and excess air on about 3-ft centers. Place the opacity meter in the stack.

No instrumentation by itself can guarantee efficiency improvement. Knowledgeable and motivated operators are essential. It is far easier for an operator to coast along at high O_2 readings with reduced risk of smoking than to run O_2 as low as possible with occasional upward bias for short periods during load transients or sootblowing. Similarly, it is easier to blow soot at routine intervals by posted schedule than to evaluate the cost of blowing soot vs loss of unit efficiency as exit-gas temperature rises. The instruments should allow operators to run as close to design efficiency as possible. Improvements beyond design efficiency usually require large capital expenditures for economizers, air preheaters, low-excess-air firing equipment, and burner controls.

W P Lauer, Jr, *E Granby, Conn*

Read temperature at boiler outlet

A stack-temperature gage can be very valuable, but a temperature element installed near the outlet will give a more accurate average measurement and can be recorded remotely. Comparing startup data with a graph of existing boiler loads vs flue-gas temperature can reveal such problems as high excess air, a dirty fireside, or short-circuiting of gases through corroded baffles or refractory seals. Any qualified instrument technician should be able to handle normal maintenance.

R W Brown, *Charlotte, NC*

6.6 Regular inspection of machines?

Our company operations include several large chemical plants, for which a central engineering department does much of the engineering planning. As an engineer in this department, I work, in part, on general policies for surveillance and overhaul of compressors, turbines, pumps, and other rotating equipment. Right now, we are sharply divided over this question: Should we dismantle rotating equipment on a periodic schedule, or should we wait until we get a clear signal that something is wrong before we decide to shut down and overhaul a machine?

My records analysis shows eight teardown inspections in the past three years, with nothing to show for the work but a big labor bill, plus charges for spare parts to replace parts damaged during disassembly. Over the same three years, we received from either instrument or human monitors over 20 shutdown signals that allowed us to make repairs in time to avert major trouble. We also had two catastrophic failures—a small turbine and a fan—which gave no advance warning.

Although my opinion is that we gained nothing from the periodic disassemblies, I would like to get the experience and opinions of other engineers on this subject. Have POWER readers prevented serious accidents or shutdowns by periodic teardown inspections? And what is the best way to schedule a program of teardown inspection?—GLB

Test on installation, then monitor

GLB's question cannot be applied uniformly to all classes of rotating machinery. From an operating standpoint, machinery critical to continued plant operation and lacking installed spares should get regular inspection. Cue the timing to outage schedules or seasonal load demands.

To get a base for knowledge of operating characteristics, test all rotating machinery soon after installation. Include flows, pressures, temperatures, and vibration levels. Monitor them at regular intervals to detect change. Proper documentation can establish a time interval for preventive maintenance before a significant drop occurs.

Manufacturers have set limits on temperature, so a good monitoring system will give ample notice. Vibration may have similarly set limits, but reaction to noticeable change in vibration readings, at unchanged operating conditions, is important, no matter if the set limits have not yet been reached. Here are two basic rules:

■ Don't disturb or shut down rotating machinery operating at normal performance level with no unusual symptoms merely for "routine maintenance."

■ Wear or problems stemming from operation will be more significant and frequent as machine speed increases.

W L DORNAUS, *Lafayette, Calif*

Decide on basis of machine type

Periodic teardown of rotating machinery is a controversial subject. On-line monitoring can give two specific benefits:

■ Warning of serious malfunction during operation, allowing action to prevent catastrophe.

■ Indication of deviation from normal operation, setting the timetable for teardown inspection and centering it on specific areas in the machine.

Well-maintained monitoring equipment has stretched out time between overhauls in many plants. A key point to consider is the type of rotating machinery being inspected, of course. Reciprocating compressors are, by nature, harder to instrument, and they consume or destroy their sealing systems. The gas handled contributes to deterioration, too, so regular inspection is almost mandatory.

Centrifugal compressors, steam turbines, and centrifugal pumps are easier to instrument and can run for many thousands of hours between teardown inspections, depending on fluid or motive gas. Teardown inspection of major steam turbines and centrifugal compressors should be performed only when monitoring equipment indicates an internal problem.

There is one Catch 22 in this posture—some rotating equipment has idiosyncrasies not normally detected by instrumentation. Incipient failures will occur, but the knowledge gained on previous teardowns is the basis for setting the future inspection dates and correcting machine trouble before failure can take place.

B F McLaughlin, *Randolph, NJ*

Make it an annual schedule—and less

As a plant engineer, responsible for a preventive-maintenance program for five years, I would advise GLB to open rotating machines on a regular inspection schedule. A general overhaul with parts replacement is usually on an annual basis, but a quarterly audiovisual inspection is a must to show need for possible overhaul. This inspection covers a checklist of critical items. If instruments check possible failure points, the inspection can be monthly. If breakdowns occur between overhauls, cut the overhaul schedule to semiannual.

A H Khan, *Minneapolis, Minn*

On pumps, wait for solid evidence before opening

"Don't open; let them run" should be the cardinal rule in all maintenance departments when it comes to centrifugal pumps. Don't open a pump for inspection unless evidence (factual or circumstantial) indicates that overhaul is necessary. Factual evidence includes a falloff in pump performance that will justify renewing internal clearances. Noise, driver overload, or excessive vibration is other factual evidence.

Circumstantial evidence includes data accumulated on the pump in question or on similar equipment. If a pump can operate 50,000 hours between overhauls, let similar pumps on similar service run that long before inspection. Corrosion causes one exception to this rule: If a pump on severe corrosive service has given short life, then, when materials are upgraded, it is logical to open the pump at reasonable intervals until you believe that a new "experience pattern" has been established.

Effect of internal wear on pump performance is easy to detect. As running clearances increase with wear, more of the pump capacity at any total head recirculates through the running clearances to a lower-pressure region.

An accurate test of the pump's head/capacity curve will indicate internal condition. Running a complete test is less expensive than opening for inspection and doesn't require taking the pump out of service. If full head/capacity tests are not practical or sufficiently accurate, an operator can train himself to detect power increases for a given flow and then to evaluate the wear at the running clearances.

Regular checks are advisable, of course. Beside six-month checks of alignment, bearing lubricant, and packing, make a more-complete one every year. Inspect bearings, repack stuffing boxes, and check vertical movement of shaft. Restore bearing clearances if they are over 150% of original. Remember, too, that a feeling of confidence is more important than opening the pump a few more times than necessary.

I J KARASSIK, *Mountainside, NJ*

Expect a warning in most cases before breakdown

Experience with hydroelectric plants, and previously with steam plants, has taught me that equipment in most cases will give some warning before a breakdown. In plants that have had annual overhauls, I have had to take equipment out of service a short time after the annual inspection, to make repairs. New equipment should get an inspection after one year of operation and another after two more years. Experience will then indicate, depending on service, either a three- or a five-year interval for overhaul with disassembly. Equipment 10 years old or more should be shut down for inspection every two to three years, with disassembly every four years.

T W HAMMOND, *Los Banos, Calif*

Let maintenance give overhaul interval

Our experience indicates that a planned maintenance program, augmented and modified by good engineering practice and failure analysis, is the best system. The major drawback to reaction maintenance is that it cannot be planned, and may result in significant damage to rotating equipment. Manufacturers' recommendations for the equipment will be the basis for a periodic-maintenance plan. Lump inspections together to allow periodic shutdown of various processes or sections of the plant for inspection and refurbishment as required.

Maintenance records are a must! Information gained on particular equipment lets you extend or shorten the inspection period on a "time to overhaul" basis rather than crisis repairs or routine "show-nothing" inspections. For help

in failure data analysis and to get early-warning signals of uncontrollable catastrophic failure, put monitoring instrumentation on every piece of critical rotating equipment to provide a history of performance. Of particular interest
are vibration pickups in two or three planes at the bearing mounts, motor
phase-current readings, and temperature readings on bearings, oil reservoirs,
and motor windings. This gives baseline information to go into equipment logs
and, eventually, into your failure-analysis program.

Monitoring these vital parameters in conjunction with your initial inspection program will give a real-time history of equipment degradation vs instrument reading or change in reading. Out of this can come an "early-
warning" system to predict impending failures far enough in advance to allow
planned shutdown, instead of mere reaction to costly catastrophic failures.

N R STOLZENBERG, *W Suffield, Conn*

Plant type dictates type of maintenance

GLB's records indicate a top-notch monitoring program. Our experience supports GLB in determining that teardown maintenance is not an effective program without establishment of reasonable parameters on which to base the
teardown. With a record of 20 shutdown signals against no productive teardowns in eight attempts, GLB's evidence supports his position that periodic
teardown is not productive. He does not indicate the vital followup to determine cause on the two "unannounced" failures, however. This investigation
would allow a more complete monitoring program.

There is little to suggest that mere passage of time requires equipment overhaul. Such factors as wear, fatigue, and deterioration need to be tempered by
the number of starts, duration of no-load operation, etc. The type of plant
operation also dictates type of maintenance program. For instance, a sugar
refinery needs its equipment in top condition before the annual campaign.
Plants with backup machinery, such as three half-size pumps, should perform
maintenance on an as-indicated basis, determined by reduced performance,
high vibration, or machine history based on output or other appropriate factors. And don't forget the importance of good operators, who will give valuable
insight into equipment condition based on their knowledge of the normal operation of the machinery.

D A TAYLOR, *White Bear Lake, Minn*

Preventive-maintenance program is working

If the periodic inspections had been varied over the years, with a gradual increase or decrease in approximately six-month intervals, GLB could have
developed some trend analyses over the equipment's operating cycle. The increase or decrease in inspection interval would depend on condition of parts
found during inspections. Trend analysis, inspection findings, and the equipment supplier's projected parts life should yield the data needed to develop a
comprehensive PMP (preventive maintenance program).

Quarterly vibration checks will complement this study. A displacement pickup could monitor compressor piston-ring wear, for instance. A signature analysis should follow return to service of a piece of equipment, to establish baseline conditions. We have put in a program similar to this. Although the program is only two years old, it has sold upper management on its worth. GLB should remember, however, that no matter how extensive a PMP he has, he will always encounter the "one that got away."

N M TERRELL, *Chicago, Ill*

Reciprocating compressors benefit from PM

In our plant we have six reciprocating compressors—three on air service and three on 500-ton ammonia refrigeration units. These machines, 35 years old, have a scheduled teardown inspection about every three years. Although this preventive maintenance has not prevented two emergency shutdowns for major repairs in the past six years, we think that preventive maintenance should be considered an insurance premium against lost production and loss of life.

As a general rule, however, scheduled teardown inspections should be cut to a bare minimum. Decision to place a machine on this type of schedule is based on not only past maintenance records, but also on data from sources such as:

- Equipment suppliers' advice
- Insurance company recommendations and case histories
- National Safety Council reports

In addition, the various plants should evaluate the hazards in each production unit in terms of maximum probable losses and degree of risk involved. A regular interplant exchange of maintenance reports and a pooling of spare parts should be company policy, too.

D G LAW, *Shawinigan, Que, Canada*

Try predictive-maintenance programs

Business economics should decide. Does it cost the company more to make PM teardowns than to accept catastrophic equipment failures? From GLB's data, the unscheduled-failure rate is about 7%. If only two of the teardowns prevented later emergency repairs, the emergency-repair rate would otherwise have gone to about 13%, intolerable to a maintenance-cost-conscious management.

GLB's question leads to the area of predictive maintenance, which is based on recorded experience or equipment analyses. Predictive-maintenance programs have worked, as I know from experience. A minimum of emergency repairs indicates their effectiveness.

B B BROWN, *Hagerstown, Md*

Thrust movement is good indicator

Installing instruments to check vibration, and especially thrust movement, should get consideration, with turbines in particular. Along with this, oil analysis is a reliable check. Look for changes in the oil, along with the presence of bearing materials and liquids. As long as thrust, vibration, power requirements, and oil analyses are consistent and unchanging, the machine is in order and should be left alone.

C R KLEIN, *Wauwatosa, Wis*

Periodic visual methods can help

Periodic visual inspections, with inspection covers opened, can be very useful, but generally a teardown of equipment that is operating without any apparent problem will do more harm than good. Record of hours of operation and repairs made will help establish a frequency for certain repairs. Regular teardowns, in my experience, give nothing but labor cost and damaged parts—and even headaches on occasion.

J BRANDT, *Hudsonville, Mich*

6.7 Assurance of thermocouple accuracy

Last week, I ran into a puzzling problem. While out in the process area of our plant with a new engineer, I was explaining some of the operations to him in response to his questions. I told him we measure temperature by thermocouples in many processes, and he wanted to know how we can be sure that the temperature reading is right. I couldn't answer this and, after checking into matters, I wonder how thermocouples can really be verified in service. We rely on them for accurate quality control—yet we have no reliable easy way to calibrate them.

Like many other plants, I suppose, we buy these devices already calibrated for the necessary temperature range, and then we install them and forget them until something goes wrong. It seems to me that this is working in the wrong order. We should be able to find out that the thermocouple is out of calibration before we get a lot of rejected product. Right now, we can't be sure that a thermocouple hasn't drifted, or, assuming that it is in error, that today's reading is close to the erroneous reading of yesterday or last week, which produced good product.

What has been the experience of POWER readers with thermocouple accuracy and calibration? Does the thermowell design have any influence on the problem? How can we improve interchangeability and reliability of our thermocouples?—FCD

Loose connections cause many wrong readouts

A thermocouple (T/C) is by nature very reliable. If checked in an ice bath or boiling-water bath, and in a calibrated oven if service range is over 250F, it should perform well for years. Our experience shows the major causes of wrong readouts, in order of frequency, to be:

- Loose connections at the T/C junction, readout meter, or intermediate junctions (about 55% of cases)
- Readout-equipment malfunctions (about 25%)
- Frayed leads, causing grounds, wrong indications (about 15%)
- Faulty T/C (fewer than 5%)

The most common problems develop in the system from the T/C to the readout, or at the readout. We find the following measures effective in detecting problems:

- Ensure material compatibility through the system.
- Before installing T/Cs, check them at three or four points, and do this in an instrument lab.
- At point of installation, verify readout of ambient temperature and body temperature, and check for upward tracking when T/C is subjected to the momentary heat of a match.
- Before connecting the T/Cs, verify system readout by applying voltages at the T/C with a pot.
- Install only three-lead T/Cs in all systems. In a critical or hard-to-read area, install dual-element T/Cs, with individual wire runs to independent readouts.
- Where thermowells are needed, make sure the ends are in a representative fluid zone and that the T/C touches the end of the well. Mark the T/C where it touches the well, so that, if it is taken out, it can be reinstalled to right depth.
- Make system check at three-to-six-month intervals to verify indication.

N R STOLZENBERG, *Suffield, Conn*

Check at least three points in range

If FCD has millivolt potentiometers, temperature baths, precision glass thermometers, and spare thermocouples, the answer is not too difficult. If all his T/Cs are in thermowells, the job is easier. First step is to make a spare T/C your reference. Check it at 20%, 50%, and 80% of range, using the potentiometer, temperature bath, and precision thermometer. Junction and thermometer bulb must be close together in bath.

With transducers, FCD can apply a millivolt signal, equivalent to process temperature, to the transducer input. If the readout indicates the correct temperature, while the original T/C gave a suspect reading, then either the T/C is bad or the process is out of limits. Substitute the reference T/C for the original and compare millivolt outputs to decide. Make sure the T/Cs are bottomed out in the thermowells, however.

G K EBO, *Ingleside, Tex*

Aluminum foil can boost conduction

Aging has little effect on thermocouples. Because they are two dissimilar metals, joined and sealed, the emf from a given temperature will be constant as long as the T/C is not exposed to extreme overrange. Calibration of T/Cs is in three steps. First, check by ohmmeter to determine continuity. Next, calibrate the input bridge for the right range span. Finally, reconnect and make a range-zero adjustment.

Most T/Cs are nonlinear if the range is wide—say several hundred degrees. If possible, however, the T/C should be exposed to a known temperature, and the input bridge zero adjusted so that, when the line shuts down for a time, the T/C will read ambient temperature. Even though process temperature may be far above ambient, there will often be an emf output that can give the right indication with correction.

Thermowells have little effect on overall accuracy, but they do cause a time lag. Sensing time can be reduced in two ways. A silicone grease or oil, or other conductor, is one way, but thin corrugated aluminum can also provide many points of contact for conduction between bulb and well.

The T/C industry complies closely with NBS standards, and a comparison of T/Cs made by different companies will show that the emf from each is normally within 5%. Interchangeability is not often a need, because the T/C is dependable and long-lasting. Reliability can be bettered by a thermowell of proper length and diameter, located away from high process turbulence. High velocity past a thermowell may cause a temperature error from friction.

J E JANDA, *Pottstown, Penn*

Solder pot is aid in calibration

We have about 700 T/Cs in use constantly, and require an accurate T/C calibration procedure. We have two setups, both controlling the heating element by a controller. One element is an electric iron; this is for surface-type T/Cs. The other element is a solder pot, serving well-type T/Cs (see sketch). We can read millivolts directly from the T/C under test, or we can connect the T/C to a calibrated recorder.

C STOVAL, *Coushatta, La*

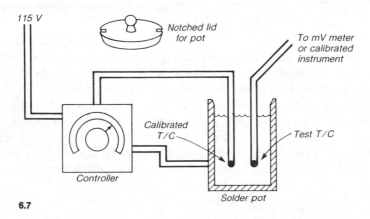

6.7

115 V

Notched lid for pot

To mV meter or calibrated instrument

Calibrated T/C

Test T/C

Controller

Solder pot

Consider accurate portable aids

If FCD's plant uses many thermocouples, it might be able to justify its own standards lab, recalibrating T/Cs periodically on a rotation system. We have generally recalibrated on a one-month frequency for devices in the 1000° C region. Judgment on this must come from a plant's own experience. The emf at the device junction *will* change with aging at elevated temperatures. If the devices must remain in the process line, portable calibration instruments can read with accuracies of ±0.1° C to ±0.01°C. Millivolt potentiometers, pyrometers, and digital thermometers are some of them. In all cases, the device is calibrated against a standard device, traceable to the National Bureau of Standards.

N R WENKO, *Clinton Corners, NY*

How about T/C wire uniformity?

FCD could ask his thermocouple vendor to manfacture all T/Cs from the same wire batch. Then a batch sample could be calibrated, and the corrections applied to all the T/Cs.

E L GRIMES, *Aberdeen, Ohio*

Alloys give sharp melting point

For moderate temperatures, try fusible alloys with sharply defined known melting points. With the T/C in the hot molten alloy, read temperature at solidification on slow cooling and again as the alloy melts during reheating.

L A BUCHALTER, *West Point, NY*

6.8 Tips on measuring stack pollution

A government air-pollution inspector checked emissions from our plant recently, and told us that if we couldn't clean up the stack plume, our 20-yr-old stoker-fired boilers would have to be shut down until we could make modifications. I don't know how we can be doing such a bad job of cleanup: We have a cyclone collector.

This whole thing of air-pollution regulations is very confusing to me. There are national standards, state standards, local ordinances, etc, but it is extremely difficult to find someone who knows enough about each to tell me what has to be done. For example, I thought all I had to do was observe stack opacity with POWER's MicroRingelmann chart, and if it was below No. 2 (40% opacity), our plant would not be considered a polluter. But the inspector said this wasn't necessarily true, and began spouting regulation after regulation, confusing me more. Of course, he would not recommend the test equipment we should use, which compounds the problem.

How can I find out what regulations pertain to our plant? What method and equipment should I use to find the particulate loading? How can I find out if the cyclone collector is doing its job properly? Should we install additional cleanup equipment? —KM

Sampler finds grain loadings

It is true that some air-pollution regulations appear to be comfusing, because in some areas it is the particulate grain loading that is important, while in other areas the emphasis is on SO_2 or NO_x. The check that KM refers to appears to have been a visual check. The state Dept. of Environmental Protection can supply a copy of its regulations and acceptable methods, and this would be the first place to turn.

For particulate grain loading, which seems to be the problem here, a high-volume sampler is advisable. The results can be obtained reasonably quickly and cheaply by semitrained operators. A test takes only about an hour, plus time for weighing and calculations. The same sampler can serve to check the efficiency of KM's collector and to evaluate the effect of coal treatment in reducing particulate emissions.

There are many stack-testing firms offering services for modest fees, too. They are familiar with regulations in the area, and will either test for the plant or install equipment and train plant operators in its use.

R P BENNETT, *Whippany, NJ*

Look into bag filters and wet scrubbers

Even though Ringelmann charts have been used for assessing smoke emissions from stacks, the pollution-control regulations in force set maximum limits on paticulate emissions, SO_2 emissions and NO_x emissions. Since KM's plant seems to be of small size, the NO_x limitations do not seem to apply. Indication of particulate emissions can be obtained by a reflected-light dust monitor or by a photoelectric sensor. The instrument can be set to initiate an alarm if the reading exceeds a preset limit. Fabric bag filters following the cyclone separator could be a solution in the smaller-size plant. High SO_2 content calls for a fuel change. Another alternative might be wet scrubbers after the cyclone separators, to reduce particulates and simultaneously absorb SO_2.

T M PURAM, *Bombay, India*

Consulting lab can sample, using EPA method

Specific emissions are regulated on these levels:
- Federal (Environmental Protection Agency).
- State.
- Air-pollution-control district.
- County.
- Municipality.

KM should start on the lowest level and work upward if he wants to find the specific regulations with which his plant must comply. Depending on his locality, the last-mentioned three regulatory levels may not exist. Every state, however, has some kind of agency for standards and enforcement.

Like the regulations themselves, approved measurement techniques vary widely. Techniques range in complexity from on-line opacity meters to simple Ringelmann-chart comparisons. Probably the most widespread method for particulate measurement is that called "EPA Method 5." The method, and the equipment required, are detailed in the Federal Register, Volume 36, No. 247, Thursday, Dec 23, 1971, under the title, "Standards of performance for new stationary sources."

KM can get a copy from the Government Printing Office or in most large public libraries. Many firms manufacture the equipment necessary for the measurements. Other techniques for sampling exist, such as those recommended by ASTM and ASME, and also some locally used methods, but Method 5 is being adopted in many states.

If KM does not want to buy his own equipment for sampling, most areas have consulting laboratories that can sample for him. Instead of trying to diagnose his cyclone collector by long-distance, I suggest that he or a consultant take particulate samples both upstream and downstream of the cyclone. Comparison of the weights of the two samples will tell him if he is currently complying with regulations. This assumes, of course, that he will use the test method specified.

J N Davis, *Mountain View, Calif*

Cyclone collector may need help

KM should know that even a thorough knowledge of regulations will probably not help him if he relies on a cyclone collector as his only tool for collecting flyash. I suggest that he look into electrostatic precipitation, in addition to instruments for finding particulate loading. Many older plants are going to have to install such equipment before they can hope to pass the tightening inspection requirements in urban areas.

R V Sorenson, *Philadelphia, Penn*

Don't confuse opacity and mass when monitoring

Before KM goes out and buys particulate-measuring equipment, he should have some idea of what he really wants to find out. His primary job is not that of scientific research, but only the passing of an inspection. Some particulate-measuring equipment is more oriented toward scientific work than to continuous monitoring of the stack gases to make sure particulate levels do not exceed the legal maximum.

Remember that there are two kinds of particulate measurements: One determines opacity; the other gives the amount of mass in the gas flow. There is no real correlation between the two, either, and this fact is the source of more and more trouble for the practical plant operator. Up to now, opacity measurements, or darkness of stack emission, were the governing ones. KM's Ringelmann numbers were the most-relied-on indicators of how dark emission was. There has been so much trouble with judgment of Ringelmann number, however, and so much money is at stake, that many authorities are going over to measuring the mass flow of particles.

Measuring the mass flow of particles can be done at night, or when the stack plume is not black in color. The latter condition has been troublesome in the past for the ordinary Ringelmann measurement. For collection of particles, filtration and weighing is often done. The size breakdown is not important, unless KM wants to determine the efficiency of his separator. Total mass emitted per hour is what counts.

Several instruments have appeared on the market for the type of sampling needed. Isokinetic sampling is necessary in this work, and the term is often heard. It means that the velocity of the stack gas entering the sampling probe is the same as the velocity of the stream passing the probe. If the velocities are not the same, then the fractions of small and large particles will be different in the sample and the main stream. The difference is more important for particles over 3-micron diameter, which are in the important range for air pollution.

For opacity, some automatic equipment is available to measure inside a stack continuously. A light source on one side of the stack sends a beam across to a photocell on the other. Lens systems direct the beam and focus it. Advantage of this method is that measurements can go on at night, and without an operator who has been trained to judge opacity. This last need was one good reason why the old Ringelmann method has lost some ground in antipollution efforts.

J T WALLINGER, *Los Angeles, Calif*

Check the cyclone's loading via emission opacity

Information on what regulations apply to a particular plant is obtainable from the air-pollution-control board or similar agency in the state where the boiler is installed. In most states, opacity of 25% or less is allowed for stack emissions. Along with opacity, a particulate loading is specified in the regulations. According to federal standards, the particulate loading limit is 65 $\mu g/m^3$. For particulate loading, an opacity meter giving opacity in percent is available. This opacity value can be correlated with particulate loading.

Before testing the cyclone collector, KM should make sure that the opacity of emissions is below the regulation limits by control of the following factors:

■ Furnace draft. Excessive draft will cause heavy particulate loading.

■ Fines in coal. Reduce fines in the crushed coal being fired. If coal has been stacked outdoors for long, the weathering effects make it more susceptible to powdering in the crusher. More fines cause higher particulate loading in flue gases and hinder proper distribution of combustion air to the coal bed on the stoker.

If changes in draft and fines fail to reduce opacity of flue gases leaving the stack, the cyclone collector should be tested by measuring particulate loading at inlet and outlet. The manufacturer of the collector can give information on method and instrumentation needed. Collector specifications for inlet loading and gas volume should be checked, too.

M SINGH, *Chicago, Ill*

Recommends starting at state-authority level

The state level is the best one to try if you want information on air pollution, we have found. This is the source of the most reliable advice, even in regard to local air-pollution authorities. If KM is in a large city, of course, there are added requirements, but he will still be wise to check with the state.

L D RITENOUR, *Louisville, Ky*

Optical-transmission photometer works

KM could try an optical-transmission photometer in his stack for measuring the density of smoke. We have used such a device and have obtained good results, after a short period of difficulty with adherent dust on the apertures. The amount of light transmittance can be related to Ringelmann number.

C P RELITSCH, *Indianapolis, Ind*

6.9 Group controls and package boilers

Five package-type oil-fired steam generators, ranging from 8000 to 22,000 lb/hr capacity and supplying saturated steam at 140 psig, are scattered all over the basement of our main building. Each boiler faces a different way, because of installation problems that I am told existed when the units were purchased, one by one, over a period of 10 years. As plant utilities engineer, I have been trying to clear up some of the space problems in the boiler room. I now am able to look forward to realigning the boilers to face a central floor, or to bring all the controls together at a central point. If I centralize the controls without moving at least two of the boilers, it will not be possible to see all the boiler fronts from any possible control station.

My estimates, with much of the work being done by ourselves, run to about $31,000 for moving the boilers, and about half that for centralizing the controls. I believe that we should do one or the other, but I can't get anyone else around here to see things my way. The owners tell me that $31,000 is too much money to spend for a boiler plant with only 84,000 lb/hr capacity. Can POWER readers give me any good management-convincing reasons why we should rearrange our boilers or controls into an orderly and neat setup? What favorable experiences have POWER readers had who have carried through with similar projects? —TL

Show a labor saving first, like one man per shift

In general, the only way to convince management to relocate boilers and/or controls is to demonstrate that a reduction in the operating labor force is possible. In my own experience with a situation similar to TL's, we found that one extra operator was needed on each shift because of a helter-skelter arrangement of boilers. Sole reason for the extra man was to provide coverage if an operator had to give all his attention to a problem in one boiler, rather than the normal tending of the two installed. For a four-shift operation, the potential annual saving in labor is about $50,000 for one man per shift. As for moving the boilers, TL will not gain much. If critical operating parameters are annunciated and all boilers equipped with flame-safety apparatus, it is not imperative for the operator to be able to see all boiler fronts from one spot.

T M DOBIESZ, *Livonia, Mich*

'First-out' alarm spots trouble fast

TL should centralize his controls. This will serve a better purpose than will setting up a neat firing aisle, even though the latter is nice to look at and would have been cheaper to install originally, provided the space existed. Overall, operation and supervision are much easier and better if all controls, indicators, integrators, and alarms are in one place.

To convince management, TL might point out that centralizing and updating controls could increase operating efficiency and might even cover the cost of renovation. With his plant operating at full capacity of 84,000 lb/hr, TL's fuel cost is probably about $125/hr. At 1% increase in efficiency would pay for the changes in 500 operating days—less than two years. Also, be sure to stress the safety and dependability angle. With the inclusion of a common alarm board having a "first-out" feature, TL can locate trouble faster.

J LAPPIN, *Toms River, NJ*

Check on how steam is used, too

TL does not give much data on the operation of his plant. What about average steam loads, peak steam loads, oil type, and whether the boilers work one, two, or three shifts? An 84,000-lb/hr boiler house is not a small and inexpensive operation, nor is its annual cost an insignificant amount. Many small industries, however, do not pay much attention to their boiler houses, and tend to lump everything together into a maintenance budget. I believe TL is looking at his boiler house from the wrong end. Unless he has serious maintenance problems because of boiler location, it is not important that he be able to see the fronts of the boilers. The $31,000 expenditure would be solely for esthetics.

Centralizing the controls is doubtless a good move. If they are used and monitored properly, TL can likely gain a couple of percentage points in efficiency, which will mean $7000/yr. This is a good return, if TL takes full advantage of centralization. The assumption is that the control panels tell him something. Instantaneous fuel flow, steam flow, burner pressure, and oil temperature are all critical data for efficiency. Controls are only tools, which must be monitored and tuned. Each boiler will have its own characteristics. The right boiler must be selected at the right time for the load at that moment.

TL's boilers are probably firetube, with best efficiency occurring between 40% and 80% of full load. Efficiency falls off quickly if the burner tips carbon up or if the breeching soots up. A check or two per shift helps prevent this, but the optimizing of performance is based on good data. A good steam flowmeter for each boiler is a must. Accurate, viscosity-insensitive fuel flowmeters are critical.

Existing controls are adequate only if there is true proportioning of fuel and air to load. The controller must be tunable and have reset functions to compensate for load change in either steam flow or auxiliary equipment. If more than one boiler is on line at a time, the controls should be cascaded from a master pressure controller. Without this, cycling and load swapping can occur and be hard to detect.

Small boilers on a common header will always fight each other to supply steam to a small, low-volume steam-header system. Master pressure control and base loading, either manual or automatic, are usually the cure for this. The centralized panel should tell TL at a glance his evaporation efficiency, fuel flow, and steam flow. With proper layout, an increase of up to 5% is not hard to get. Once a company becomes energy-dollar conscious, further savings can come from study on how the steam is used. This is where the big money is; up to 30% of the steam could be wasted.

D PERRY, *Philmont, NY*

Install a monitoring system, floor space permitting

The estimate of $31,000 seems somewhat high. A more practical solution for TL might be the installation of a monitoring system, such as those made by several firms. The present usable floor space should be checked for the possibility of installing one of these systems. Once installed, the system will allow the operators to devote more time to taking care of the boilers instead of running from boiler to boiler to check on air and other variables. The savings could pay for the installation in two years or so of use.

J V MUNSON, *Spring Valley, NY*

Move controls instead of boilers to central location

It would be a big advantage to have all the boiler controls centralized in a location equidistant from TL's three largest boilers. This location would put the control panel close to 75% of the steam source. Controls for such auxiliary equipment as feedwater pumps, feedwater heater, and oil pumps should also be there. The boilers need not be located centrally if the controls and alarms are adequate to monitor the boilers and to shut them down.

J BRANDT, *Hudsonville, Mich*

7 Energy Management

7.1 Guidelines for waste-fuel evaluation

Over the past 10 years, several of the smaller plants in our company have had opportunities to burn waste materials of highly varied composition. Some of the boilers have had all kinds of stuff put into them that could cause trouble from either environmental or corrosion standpoints. In other plants, refuse that might be acceptable has been turned down. I would like to draw up guidelines for the plants to assure that we will utilize energy opportunities without harming our boilers or bringing environmental authorities down on us.

We have waste lubricants, packing materials, knit-goods remnants, and grass and brush cuttings, for example. What we need are general and easy-to-understand rules for deciding what to burn, how to segregate at low cost, and how to burn the waste without reducing load on boilers or risking damage to grates, boilers, or flue passages. What have Power readers found to be the best and most concise rules for selecting waste fuels? What are the tests we should make beforehand? Can we rule out ahead of time entire classes of refuse and waste? Can the decision be left up to the individual plants? — LC

Match your waste and primary fuels carefully

We burn our process-generated waste continuously in the boilers and occasionally burn outside waste fuels. The first thing we ensure is that the waste liquid is compatible with the fuel in use. It should not alter its viscosity considerably upon mixing. Viscosity of the mixture cannot be too high or there will be operating difficulties. Also, fuels with heating values substantially lower than our primary fuels are rejected, because they cause flame-stability and control problems. If LC is anticipating only small quantities of waste fuels, this may not be a problem.

Generally, we baseload with liquid waste and modulate steam demand with the primary fuel—in our case, natural gas. Our boilers are packaged units with conventional feedback control. Instruct operators to make small changes in baseloaded or secondary fuel, especially if only feedback control is used.

G C SHAH, *Houston, Tex*

Complete burning is important

Small steam generators fitted with traveling-grate or agitation stokers can be adapted straightforwardly for woodwaste, wood chips, bark, sawdust, shredded paper, farm silage, grain, and nut shells. For other wastes, make sure that the material will burn to completion, that the products of combustion are not toxic or corrosive, and that the unburned debris will pass through the combustion zone to the ashpit. For example, waste greases and heavy lubricants must be leadfree. Used motor oils, polyvinyl chloride materials, and lead-base paint refuse should generally be avoided because they emit toxic vapors. On the other hand, units that burn atomized fuel oil can be adapted for contaminated turbine oils.

Heavy metal objects and glass should be preseparated and sold as scrap. Highly combustible material—such as shredded tires—is best fed intermittently with less volatile materials to avoid smoke and obnoxious odors. Winter leaves can be burned directly, but wet fresh-cut grasses should be burned with the primary fuel directly on the grate section. Avoid high air turbulence at the grate surface, because this will lift unburned material from the combustion zone and deposit it in the gas passages, potentially causing secondary combustion and/or pluggage.

C B WILLIS, *Livingston, NJ*

Talk to manufacturers before deciding to burn

Before deciding to burn waste products, LC should first discuss the situation with the boiler and auxiliary-component manufacturers. Allowable heat-release rates, fuel-ash fusion temperature, minimum tube spacing, and refractory spalling characteristics all influence performance, and the degree can best be gaged by the manufacturers' engineering departments. They can also help determine whether the fuel savings will justify any modifications to the boiler and waste-fuel-preparation, environmental-control, and ash-handling equipment.

The first item they will surely want to see is an analysis of the waste material, including: low heating value, ignition temperature, volatile matter, fixed carbon, ash content, ash fusion temperature, hydrogen content, sulfur content, moisture levels, explosiveness, toxicity, odor, corrosiveness, and lube-oil sediment content, if that waste is being considered. Some particular points to be aware of:

■ Heating value and combustion efficiency of different waste products vary appreciably. Unless large quantities are available for long periods, frequent modification may be required of the auxiliary-fuel firing rates, combustion-control systems, and ash-handling systems. Keep in mind the labor and maintenance costs.

■ Because LC is responsible for many boilers, it may prove advantageous to collect the material from all the plants, then burn it at only one location. The reduced capital, operating, and maintenance expenses may defray the greater transportation and storage costs.

■ Large quantities of combustible ash generally result from burning waste products separately or with conventional fuels. To avoid ignition conditions when this ash accumulates in flue-gas ducts, chimneys, and hoppers, the system may require higher flue-gas velocities and intermittent operation of the ash-removal equipment   H B Wayne, *Jamaica, NY*

Try for corporate commitment

Using waste fuels is desirable from the standpoint of US energy policy and conservation practices. Efforts to encourage this, however, still must be justified economically for management acceptance, because waste-fuel combustion is desirable but not necessary. Therefore, I believe it should be made corporate policy and not left up to the individual-plant management. Plants should be required to perform the necessary engineering work but not to fund it as direct costs. Instead, the charges ought to be incurred by the central corporate treasury as a corporate energy-conservation effort.

I would further not recommend any ruling out of an entire class of wastes without preliminary tests. One important consideration is whether various combinations of wastes can be simultaneously burned without undesirable products of combustion or operating conditions that might not appear when each is burned separately.

B B Brown, *Hagerstown, Md*

With big companies, it's a plant-by-plant proposition

This can be a very ticklish problem, if my experience with a multiplant outfit is any indication. We had several different products, and each plant had waste material peculiar to its own location and operation. There was much initial footdragging when company management decided to improve the boiler cost picture by burning waste, but gradually the separate plants came around to grudging cooperation. We never got much more than that, in spite of general agreement that putting waste materials to work as fuel was a good idea.

Our solution was to decide plant by plant just what wastes would be burned and then to leave it up to the plant to get the task done. This worked much better than the detailed combustion instructions that I have seen issued by some firms—instructions that are rarely followed, because conditions change from time to time. One thing we never attempted to do is to burn any fuel that could give off a toxic waste in either flue gas or solid residue. This meant segregating some packaging materials, but the labor cost simply had to be accepted as part of the fuel-preparation charges.

M F Carey, *Milwaukee, Wis*

Shredding may be necessary for solid wastes

From the term "smaller plants" in the problem statement, I assume that I C has either packaged oil- or gas-fired boilers, or small stoker-fired units for coal. It is useless to talk about burning wastes in pulverized-coal-fired boilers, even though attempts are regularly made to do this. For an industrial boiler-plant, there is too much at risk to permit any but regular coal fuel to go into the boiler.

With gas-fired boilers, there is practically no waste that can be fired. With oil-fired boilers, most of the waste oils can be burned, but the percentage must be low, perhaps 2-4% blended in with the regular fuel after water settling. For solid waste material, LC will probably have to invest in some shredding equipment to get piece size down to where the piece will burn out during normal grate-holding time. Then the waste can be added directly to the coal and as close to the stokers as possible.   J E WALLACE, *Cincinnati, Ohio*

7.2 Figuring kW loss from pressure drop

The high cost of energy is making everyone more receptive to fuel-saving measures. Part of our effort as consultants now goes to review of system-design features formerly taken for granted. Here is a recent problem: A 35-MW turbine/generator is to be supplied by 350,000 lb/hr of 505-psig, 750F steam from two waste-heat boilers, one rated at 125,000 lb/hr and one at 225,000 lb/hr. Both boilers are to generate steam at whatever pressure increment over 505 psig is necessary to overcome the drop through piping to the turbine. From each boiler, we plan to run a 10-in. line into the end of a 10-in. tee, where the flow will combine and flow out the side through a 10-in. line to the turbine, as in sketch (7.2A).

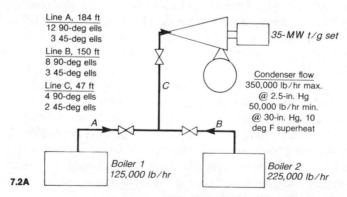

Line A, 184 ft
12 90-deg ells
3 45-deg ells

Line B, 150 ft
8 90-deg ells
3 45-deg ells

Line C, 47 ft
4 90-deg ells
2 45-deg ells

35-MW t/g set

Condenser flow
350,000 lb/hr max.
@ 2.5-in. Hg
50,000 lb/hr min.
@ 30-in. Hg, 10
deg F superheat

A   *C*   *B*

Boiler 1
125,000 lb/hr

Boiler 2
225,000 lb/hr

7.2A

We decided to calculate the loss in kW of generator output caused by the tee. So far, however, our staff cannot agree on this, nor can other experts whom we have called in. Some of us think the tee configuration may cause operational problems—but we disagree on this, too. Can POWER readers give us any clue on calculating the kW loss in the turbine/generator set? Should we expect operational problems in control, hammer, vibration, etc? And is there a better, low-loss way to combine boiler flow?—NP

Loss is estimated at 10 kW from pressure drops

The fitting pressure drops for converging unequal flow rates under the conditions given by NP are found, from published curves, to be about 1.1 psig from B to C, and about 0.1 psig from A to C. Note that diverging streams, flowing out in two directions, would give different pressure drops.

The kW loss from these pressure drops can be determined in several ways, such as derivation from noting that the turbine expansion line end point on a Mollier diagram shifts upward to the right. Because NP's turbine is a condensing one with a wet exhaust at full load; however, an entropy balance over the cycle is easier, with kW_{loss} = hourly entropy increase × condensing temperature ÷ 3413. Entropy increase is in Btu/deg F/hr, and condensing temperature is in deg F abs, so the loss will be 4.16 kW/psig/100,000 lb/hr, at 2½ in. Hg., which gives 4.16 [(2.25 × 1.1) + (1.25 × 0.1)], or 10 kW. I hope that some of NP's experts agree with me.

R E MATHYS, *Columbus, Ohio*

Should strive for more uniform drops

Based on charts, the equivalent lengths of lines are: 455 ft for A, 341 ft for B, and 148 ft for C. For 10-in. Sch 80 pipe, the pressure drops at given flows would be 5.3 psig for line A, 12.1 for B, and 11.8 for C. Changing A to 8-in. will increase its drop to 12.0 psig, consistent with 10-in. B and C. Alternatively, changing B and C to 12 in. will decrease their drops to 4.2 psig and 4.1 psig, consistent with the present 10-in. A. Line C will have the lowest drop in either case. This is necessary for good flow.

Next, the opposing flows should not go into a tee on the run. Flow A should make a 90-deg bend, and flow B should connect with it shortly downstream as a lateral entering at 45 deg. For comparative losses, the tee would give a loss of 58 ft of 10-in. pipe, and with two flows the loss would be 116 ft. Turbulence and unbalanced pressures will modify this, of course, but the result is close enough. For the proposed junction, with 10-in. pipe, the equivalent length is 52 ft. Saving is 64 ft of equivalent pipe length. Pressure drop saving is 9.25 psi–5.2 psi and the kW loss, calculated on the basis of changed specific volume of the steam, is 166 kW. This is approximate, of course, but the tee junction is certainly less efficient than the bend and 45-deg lateral.

B B BROWN, *Hagerstown, Md*

Tee loss can be slashed by larger line

The proposed piping will give a high pressure loss, especially because of the opposing flow out the side of the tee. Steam velocity is high, too. In line C, assuming Sch 80 pipe, velocity will be about 15,000 ft/min. Usual practice for lines to steam turbines is 12,000 ft/min max.

At the fitting itself, shock is the major cause of pressure loss, and turbulence from sudden changes in flow direction is the agency. Pressure drop from shock in fittings can be found from: $\Delta P = (v/12) [\kappa(G/10^5)^2]$, with ΔP the pressure drop from shock, in psig, v the average specific volume, in ft^3/lb, G the mass flow in the smallest cross-sectional area, in $lb/ft^2/hr$, and κ a shock loss factor depending on the fitting. For a tee, κ is 1.8.

Pressure loss, working with the Babcock formula and the above equation, in line C with Sch 80 pipe is 4.75 psi for 47 ft of straight pipe, 8.45 psi for the tee, 6.0 psi for four 90-deg elbows, and 2.0 psi for two 45-deg elbows, a total of 21.2 psi. The tee gives 40% of the total. To calculate power loss because of line pressure drop, first determine the available heat drop from steam tables or Mollier diagram.

It is the difference in enthalpy between steam supply at turbine throttle and at turbine exhaust. Here, the reduction in heat drop from line-C pressure loss is about 5 Btu/lb. Based on maximum steam flow, the gross power loss would be 514 kW. Net power loss will be reduced if the heat rejected to the condenser can be sent to process or district heating.

B M Kine, Vancouver, BC, Canada

Elbow and wye are good combination

Bullhead tees should be avoided, because they can give an unstable, oscillating condition, indeterminate and therefore impossible to evaluate. As I estimate it, NP's tee can be expected to have vibrational operating problems and a pressure loss equivalent to about 100 kW.

The estimate is based on total loss of steam velocities approaching the C junction from boilers A and B. Total loss of kinetic energy equals mass times velocity squared, divided by g. Values are 119.8 ft for A and 388 ft for B. Multiplying flow rate by head in feet gives ft-lb work lost, which comes to 7.56 hp for A and 51.65 hp for B. The total equals 44 kW.

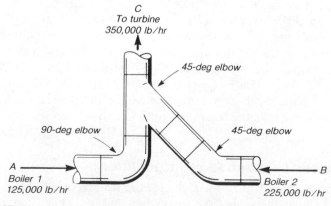

C
To turbine
350,000 lb/hr

45-deg elbow

90-deg elbow   45-deg elbow

A
Boiler 1
125,000 lb/hr

B
Boiler 2
225,000 lb/hr

7.2B

Most practical method for combining streams is an elbow plus wye combination (sketch, 7.2B). This lets the two streams form parallel venturi shapes so that they can readjust velocities to the exit-pipe velocity without permanent energy loss. A cruder method would be a baffle plate in the main header to cause the streams to change direction before combining, reducing pressure loss by at least this extent. When we realize that this amounts to square-elbow loss for each stream, we can see how inefficient the bullhead tee is—far worse than the square elbow.

The square-elbow loss is likely to be 1.3-1.5 times the approach-velocity head calculated above, so that the real loss would be over 60 kW. Actually, it is not hard to believe that the real loss with a bullhead tee would be over 100 kW, or 0.3% of the 35-MW plant output. Loss with the elbow-wye setup would be closer to 0.3 × the 44-kW velocity-head loss, or 13 kW. For best results, the wye should carry the higher flow, so that the velocity components in the flow direction will be more nearly equal, cutting differential-velocity (shear) loss.

F HASSELRIIS, *New York, NY*

3% pressure drop is reasonable

With steam-turbine requirements of 350,000 lb/hr at 505 psig and 750F, the pressure drop for the steam-turbine control valve can be evaluated as 50-75 psig. A 3% pressure drop can be allowed for the main steam piping between boilers and turbine. In this case, the boilers should operate at 600 psig and produce superheated steam at 775F. Main steam piping should be seamless carbon steel, A106 grade B (minimum), with wall of 0.500 in. All 90-deg elbows should be long-radius.

A tabulation of average velocities for all flows and practical pipe sizes is next, followed by evaluation of equivalent piping pressure drop for various sizes and flows. At the tee itself, the velocities of steam flows will vary with the steam quantities produced by each boiler, and in consequence, vibration will occur. The floating piston in the stop/check valve of each boiler can have an effect here. Condensate slugs during warmup and operation will also make their pressure known.

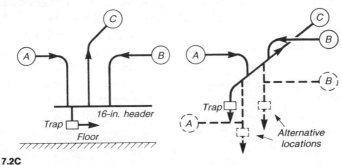

7.2C

Possible better layouts for replacement of the tee are shown in sketch, 7.2C. NP should remember that what is wanted here is to have the highest enthalpy differential for steam-turbine operation, so that the turbine can operate at optimum efficiency.

R ARSENAULT, *Longueuil, Que, Canada*

Manifold will improve stability—not tee

I suggest staying away from the tee arrangement. It would result in one boiler tending to grab load, with a result of poor pressure control and boiler instability. I recommend a manifold, which will be a plenum for distributing steam to the turbine/generator set. For control, a master pressure control loop is advised. The turbine/generator will respond to load demand by moving its inlet control valves. Pressure in the manifold will change as a function of this load demand, and the deviation from setpoint will change the individual boiler firing rate controls.

As steam flow increases, pressure drop in line C will also increase. To have a constant steam inlet pressure to the turbine at various loads, the master pressure setpoint must be varied. A total steam flow index or a generator load index will do this. This system, because of the accumulator effect of the boilers, will allow rapid load changes. Also, the bias control station will allow selection of a loading sequence as well as a means for operating the units at their most efficient point. The problem gives no details on the boiler, but NP should consult control specialists in designing a more complete control loop.

D F MADURA, *Mystic, Conn*

Header, smaller pipe help flexibility

Quite a few things appear to be wrong with this problem. First, because the boilers are waste-heat boilers, the quantity of waste heat will probably fluctuate depending on process, so NP can convert only that quantity that can be supplied steadily. Peaks will be lost. Therefore it seems of no consequence to dwell on the loss in one tee. In addition, piping arrangement must give adequate flexibility between anchored points. A 750F temperature will give 6-in. expansion in 100 ft of piping. For Line C, assuming a start at an anchored manifold, the piping with four 90-deg elbows and two 45-deg ones must consist of several short stiff legs between fittings. Result would be a 10-in. system with little flexibility and excessive thrust against the turbine connection.

Further, the turbine/generator set gets steam from both boilers simultaneously. It would therefore be wise to stay with accepted practice and avoid opposing flow at the tee. In accepted practice, boiler leads and turbine feeder line terminate at an anchored manifold, with drain leg and trap (sketch, 7.2D). A rough check on flow shows that lines A and B could be 8 in., and A could even be 6 in. and still be compatible with 10-in. line C. This reduction would help improve piping flexibility. A M PALMER, *Brooklyn, NY*

Better to increase line-C size to 12 in.

When two or more boilers furnish steam to turbines, a pipe header should be installed. A header can attenuate pulsations caused by pressure differences, and it allows connecting of future units without altering present facilities or interrupting operations. I estimate pressure drops as 13 psi for line A, 32.2 psi for line B, and 20 psi for line C. This assumes a 14- or 16-in. header instead of the tee. The drop of 52 psi through B and C represents 10% of steam pressure, which is high. The 350,000-lb/hr flow through line C would be at 19,500 ft/min, too. A minimum diameter of 12 in. would be better for C.

H B WAYNE, *Jamaica, NY*

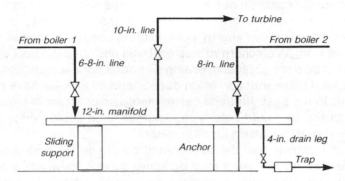

7.2D

Boiler needs attention, too, with controllers

A wye fitting would be preferable to a tee, even if two more 45-deg elbows had to be added. In the tee, the two opposing streams, at 90 ft/sec and 150 ft/sec, are almost certain to cause noise and vibration. Regarding energy savings, NP should look for a more direct piping run. A 90-deg bend is now in the lines every 15 ft on average. Loss extra over straight pipe is 200% for line C, 130% for B, and 160% for A. Rough calculation indicates outlet pressure at the tee of 505 + 12 = 517 psig. Boiler 1 will have to operate at about 517 + 6 = 521 psig to furnish its share of load, and boiler 2 will have to operate at 517 + 14 = 531 psig. This could give a control problem, although that will depend on the system operating philosophy chosen by NP.

If the turbine operates in a load-follow mode, with turbine inlet pressure controlled by changing waste-heat input, then a three-mode (PID) master pressure controller is advisable. This type takes sensing from near the turbine and provides a pressure-demand setpoint signal to each of two proportional-only boiler-pressure controllers sensing near the boilers. The proportional bands of the two controllers can be adjusted to account for differences in line losses and available energy inputs. With split ranging, limits, and different proportional bands, one boiler could be made to lead, if one waste-heat source is preferred over the others. Control stability should be easy, with conservative adjustments.

On the other hand, if the turbine is to use all it can get, little pressure control is needed. NP may want to limit reduction of inlet pressure by throttling back on the turbine, through the master pressure controller signal. The three-controller system might be wanted, too, to cover periods when waste heat could exceed electric demand.

J A NAY, *Allison Park, Penn*

7.3 How to recover condensate heat

The old question of what to do with drains from oil heaters has resurfaced in our company. It has been a long while since any of us have had experience with attempts to recover either the condensate or its heat. Recently, however, our energy-conservation committee began to study the losses from all discharged wastes. I was assigned the task of finding what could be done to reclaim heat in oil-heater condensate, and also the liquid itself. I know that corrosion of coils and oil leakage have been big problems in the past. Perhaps better materials or steam at higher pressure than the oil would help here. Another need could be for a quick-acting and sensitive alarm for oil in water.

We tend to think that the condensate should get no consideration as boiler feedwater, but there may be some service for which it would be entirely satisfactory, even if there were a little oil in it. Certain preliminary rinsing operations might benefit from this condensate. Have POWER readers had any experience with recovery of heat from condensate from fuel-oil heaters? What has been readers' experience with improved materials and design of heat exchangers for the purpose? What uses are recommended for the condensate around the steam plant? —KA

Oil indicator is a must; also sponges

Here is the result of some experience in marine installations which, with minor modifications shown, may be good for stationary plants. A two- or three-pass

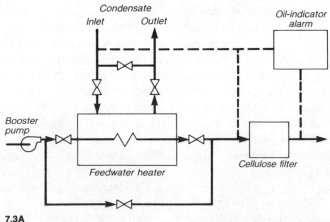

7.3A

closed feedwater heater on the makeup-water line is one possibility (sketch, 7.3A). Increase in makeup pressure to excess over condensate pressure will eliminate possibility of contamination, although a booster pump may be necessary, depending on the pressure of incoming water. An oil filter or grease extractor, downstream of the heater and just ahead of the hotwell, would give added protection. The filter can be charged with a synthetic cellulose sponge, which is cheap and has about three times the absorbing power of loofah sponges, which we formerly used. The oil-indicator alarm is a must for monitoring condensate discharge at water-heater inlet, hotwell, and filter inlet and outlet.

A slightly more expensive method relies on draining the condensate to an inspection tank, where a sensitive oil-indicator alarm monitors it and is connected to an automatic discharge and dump-valve system for release to a slop tank if the probes in the discharge lines sense oil. The clean condensate goes to a transfer pump, which delivers it to the first-stage heater through a cellulose-charged filter for extra protection.

A A CLARKE, *Jackson Heights, NY*

Try tank with float, solenoid valves

An insulated tank, with conductivity meter, float valve, solenoid valve, and audible alarm could be a way to recover condensate. The oil must be at least 10 API gravity. Condensate would enter the tank and fill it to the point where the float valve would open. Then, as long as the conductivity meter did not detect values above preset ones, the solenoid valve would be energized open to allow the condensate to pass out through the float valve and the solenoid valve in series. If oil enters the system, the conductivity meter will close the solenoid and sound an alarm. The water level will then rise and spill to a drain.

D WINCHELL, *New Britain, Conn*

Supplement by electric oil heater

If there are enough condensate drains from other sources, then a floc filter system and a condensate tank may be justifiable. Piping, valves, and pumps must be part of the system, too. Clean condensate and its heat will go back to the feedwater heater. An electric fuel-oil heater is suggested, too, for cold startups and for periods when the steam-heated fuel-oil heaters must come out of service for cleaning and maintenance.

D DUVO, *Northridge, Calif*

Compare flash tank with reboiler

The simplest way for KA to recover part of the heat lost in the condensate is to install a small flash tank and heat exchanger in the makeup-water system. Steam from the flash tank can be vented to atmosphere, or to a deaerator if conditions permit (sketch, 7.3B). Another method of safely heating fuel oil with minimum loss of condensate and heat is a closed-loop system with a reboiler as the source of heating steam (sketch, 7.3C). Steam for oil heating is generated in the reboiler, to which oil-heater condensate is returned. Primary steam to the reboiler comes from the boiler. Condensate returns to the feed-water system. Oil leakage to the closed system can be detected by periodic sampling, and the primary system is not affected.

S F Cowen, *Jacksonville, Fla*

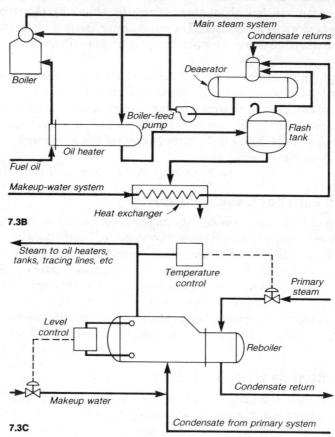

7.3B

7.3C

Go easy with conductivity meters

A better sequence of wording is "oil leakage and corrosion of coils," because the sulfur in the leaking oil reacts with the hot steam to form acid, which attacks not only coils but also tube sheets, shells, traps, and piping. The trouble with most alarms based on conductivity is that acidulated and alkaline condensates are quite conductive electrically. A pH cell, on the other hand, would become blinded quickly by an oil film and be useless. Leak detection in fuel-oil heaters is best done by alert operators or by daily sampling of condensate to seek out oil leakage. A frequent source of oil in condensate is leaking gaskets between the head, flanges, and tubesheets.

Look cautiously at setting oil-heater steam pressure higher than oil pressure, because steam condensate leaking into the tank will eventually cause corrosion failure of the bottom. There is sometimes enough water entrained with the oil to cause problems as matters stand. Economics does not favor making large investments to recover condensate heat. Tracing of No. 6 oil piping is one possible use. Residual oils flow excellently at 125F, and the condensate can trace the oil piping, recovering about 85 Btu/lb of condensate. Heating domestic water to 145F will recover about 55 Btu/lb of condensate. This assumes that condensate is available at 210F. Space or process heating and boiler makeup could give a little better recovery, say 125-140 Btu/lb.

B B Brown, *Hagerstown, Md*

7.4 Oil burners improve old boilers

When I became chief engineer for a northeastern industrial plant five years ago, I was mainly concerned with process machinery and piping, not the boiler plant. The energy crisis has forced us to give more attention to the old boilers now. Our boilers must be among the oldest water-tube ones still running in the state, but they are burning the most modern fuel—oil—in large quantities. Our efficiencies are reasonable for these types, about 70-73% as near as I can figure. I base my estimates on steam flow and oil receipts over a period. We are running flue-gas temperatures of 470-500F, with considerable excess air.

We don't have enough money to put in effective heat-recovery equipment, so we are looking at burner changes. We have come across burners that appear to give higher efficiency than the ones we now have. Some of these operate on acoustic principles. Most of the suppliers believe they can cut our fuel costs, but I am not sure. Do Power readers have any suggestions for the optimum burners for our one 65,000-lb/hr and two 45,000-lb/hr 125-psig boilers? Remember that I would like to test the results, too, so please give suggestions on this. Our load is steady; does that make our boiler-plant problem simpler?—VHR

Upgrade combustion control first

One way for VHR to solve the problem is to conduct a series of tests to determine the excess-oxygen content of combustion products. These tests should be at various boiler loadings, both within the boiler furnace and at the gas outlet. Once the data are in, calculate efficiency of the existing burner setup and the air-infiltration loss through the boiler setting. As a guide for evaluating results, most steam-atomizing burners can operate successfully at 3-4% excess oxygen levels on No. 6 oil.

Improvements in the combustion-control system are often more beneficial than burner replacement. Also, if VHR doesn't have equipment for the tests, he might consider hiring a firm specializing in this work. If test results show that the burners are the problem, then carefully evaluate low-excess-air burners. Payback period is a good way to evaluate this, provided that the existing boilers won't need major expenses soon.

R C HOGAN, *Canton, Ohio*

Watch economics via meters and analyzers for payout

VHR should start by making two purchases, even if he has to rob the company coffee fund to do it. First, buy and install some accurate oil meters. Second, get a good-quality flue-gas analyzer to measure at least the CO_2 and O_2. Then the boiler efficiency and excess air can be accurately found for the present state of the boilers. With burners adjusted for maximum efficiency, VHR will have some accurate figures.

Next, check manufacturers' data on new burners and compare the savings promised against burner installed cost, plus controls, accessories, and service requirements. If savings justify buying new burners, choose the most promising types and get names of users from manufacturers. Contact the users for information on theoretical savings and—more important—equipment reliability. Even the best of burners will not guarantee savings. Only with servicing and frequent monitoring of the combustion process, with adjustment when necessary, will optimum furnace conditions come about. With today's energy costs, once-a-year servicing and adjustment is not the answer.

J P GERVAIS, *South Porcupine, Ont, Canada*

Try the ASME short-form test first

Here are some pointers for VHR to consider before he spends money for new burners.

■ Rather than basing efficiency calculations on steam produced and oil receipts, VHR should take the more-accurate ASME short-form test. This test gives a good indication of combustion and excess-air readings.

■ Before the ASME test is run, all controls should be calibrated, so accurate readings of fuel flow, steam flow, and air flow are obtained. The fuel-control valves and air-flow control should be calibrated, too. If air-flow control is by duct dampers or inlet vortex assemblies, they must all operate to manufacturer's specification. Check burner gun tips for erosion and wear, too.

■ If the boilers are field-erected tube and tile type, instead of factory-assembled package type, and if VHR's efforts give the same efficiency of 70%, then it is unlikely that any further gain in efficiency will be possible. The high heat losses in this type of boiler prevent reaching the 80-82% efficiency achievable in the newer package types. An economizer or air preheater would give 3-5% higher efficiency. The values given for furnace-exit-gas temperature appear to indicate high heat loss, assuming that the measurement was taken right at furnace outlet.

■ Integrity of casing and refractory needs a check if the boilers are as old as VHR implies. Major refractory repairs are probably indicated.

■ The economic side of the question should be looked at. If VHR decides to install new burners, how many years will it take for them to pay for themselves? Have the boilers that much life left in them, or will retubing and major repairs be needed within two or three years? I hazard the guess that it will make more sense economically to live with the present efficiencies for three or four years, and then buy new boilers with economizers.

R A MAXWELL, *Charlotte, NC*

Is heat recovery the real answer?

Efficiency as estimated by VHR on the basis of steam flow and oil receipts does not reflect the efficiency of oil burners alone. Instead, it includes the effectiveness of utilization of heat liberated from combustion of fuel oil. With VHR's good quality fuel oil, excess air should not exceed 3-5%. If excess air is high, then either the combustion-control regulation loop is not functioning satisfactorily or the burners are not atomizing the oil properly. In the latter case, excess air is needed to get complete combustion. Proper differential pressure between atomizing medium and oil, along with cleaning and, if necessary, replacement of burner nozzles, are necessary.

VHR should check previous records to see if the high-flue-gas-temperature condition has existed since startup. If so, then the boiler design is limiting, and only installation of heat-recovery equipment will improve efficiency. If earlier flue-gas temperature was not high, then cleaning of the heat-transfer surfaces, inside and outside, can help.   R K JAIN, *New Delhi, India*

Let a specialist survey the setup

Because of the limited information available, the only viable recommendation that I can advance is to have a survey made by a reliable, qualified firing-system specialist. He might be a consulting engineer, of course, but is more likely to be an experienced contractor, burner distributor, or field representative of a qualified industrial-burner manufacturer.

Only one clue to the problem is reliable: "The boilers are very old." Exit-gas temperature is reported to be 470-500F, but is it? Is the gage accurate and properly installed? The plant "probably has considerable excess air," but how much? And what is the grade of oil? What are the CO_2 and O_2 levels? For old brick-set boilers, 70-73% efficiency may be reasonable, but are the steam-flow meters and oil receipts accurate?

Fuel costs can always be cut, of course, but the benefit may not justify the investment. This problem is so similar to the usual request for an answer without access to the pertinent data that I feel obliged to reply on behalf of engineers presented with such situations. The answer may be one of these:

■ Adjust present equipment and train operators

■ Install a new firing unit in the existing boiler

■ Replace both boiler and burner

Greatest return on cost could result from application of any of these solutions.

R C WRIGHT, *Monroe, Wis*

Excess-air control is important to combustion

The question leads me to believe that VHR presently cannot examine his combustion conditions. An orsat device or a simple CO_2-analysis device would let him to do some rudimentary investigation. This equipment is not expensive. If current conditions show stack-gas CO_2 level appreciably below that expected for complete combustion (as calculated from fuel analysis), improved atomization may boost efficiency. In the past two years, I have had very good results with the acoustic-type atomizer, and have found that combustion did improve. Remember, however, that unless you can get some control over excess air, and unless conditions involve significant stack losses of unburned fuel, there will be no chance for higher efficiency.

W E KEELS, *Channelview, Tex*

7.5 Steps in applying demand monitoring

We want to apply demand monitoring if possible in our insulation-manufacturing plant. Our plant develops substantial demand charges, but we are not sure whether they can be avoided by demand-control equipment. Sophistication of controllers is another aspect of the problem. We don't want one that is too simple, and yet an overly complicated one might cause unnecessary headaches. We would like to know what POWER readers think about:

■ How do we go about finding what our sheddable loads are, and for how long can we cut them off in a given demand period?

■ Will we have trouble with our local utility? They may not want us to take loads off and add them again continually just to avoid demand charges.

■ And what about violating local codes on worker safety and comfort? We don't want to bring OSHA down on us.

What are user experiences in this area? With a/c especially, where do we intrude on worker comfort? — LP

Classify loads three ways after analysis

The first step is a detailed load analysis by circuit, building, process, or whatever LP's appropriate unit is. The analysis should be for two months, one in the winter and one in either July or August, when air-conditioning load is highest. Next, divide uses into three classes:

■ Essential, such as continuous processes that cannot be shut down without damage to equipment or material in process, or processes with a serious start-up problem.

■ Desirable, which can be shut down without damage but at a loss of production.

■ Optional or marginal, which LP may find includes such loads as coffee makers, spotlights on the company flagpole, and the automatic door opener on the executive washroom.

The third step is to create a 24-hr job of load dispatcher. He will be in full charge, and can pull the switch on any load at any time, based on the results of the second step above. For the present, LP probably does not have to worry about OSHA. The local utility should be spoken to, since conditions differ in every part of the country. In my opinion, utilities do not impose demand charges as a profit maker but as an incentive for you to avoid peaks.

F G Norris, *Steubenville, Ohio*

Make one man responsible for program

Simplest solution to this problem might be to obtain the services of a consultant experienced in the processes to be controlled, and knowledgeable in the various peak-demand shedding techniques, hardware, and computer-software support that will reduce electrical demand cost. A general review of the demand equipment on the market will reveal a price range of $650 to $65,000, along with claims of from 15% to 45% in savings, for methods ranging from simple demand control to an entire power-management system.

In most cases, these are legitimate claims—if the hardware sales engineering and systems engineering are applied properly. Communications must be good between the plant's process engineers and the hardware vendor. The disappointments come when no one will take system responsibility, and this happens often.

To apply demand control without a well-planned and systematic review of the manufacturing process to be controlled will lead to near-chaos in the plant. In such a case, application of demand control may save some money in demand charges, but the production loss will exceed power-demand charges by tenfold —a hefty difference.

Some pointers from application of energy-management standards can come in handy when planning for demand monitoring, and here are some that I have found useful:

■ Prepare an energy and cost/benefit analysis for all environmental systems and building components and materials, as part of a schematic design that uses all architectural and engineering disciplines.

■ Consider use of smaller-sized motors to reduce peak electrical demand. Select horsepower for no more than 110% of maximum full-load requirements. Oversize motors result in low efficiency and contribute to low system power factor.

■ Schedule heatup time on equipment over an extended period, and cycle it if a number of equipment items are involved. If heatup load can precede the peak period, LP will have a good reduction there.

■ Transfer solar heat from the south, east, and west sides of buildings to north, west, and east sides at appropriate times by a common return-air plenum.

■ Look into central monitoring and control to optimize operation of all air-handling and other mechanical systems.

LP can get much free information and effort from sales engineers of many reputable companies that serve the electric-demand market, but the ultimate responsibility lies with the plant's own process engineer—and with the individual who takes total responsibility for safe and economic application of the demand device to existing plant and equipment.

P J HORVATH, *Port Chester, NY*

Sequence overlapping operations

Finding sheddable loads requires a review by the operating and engineering departments of all plant functions. Information needed is peak demand times during the day, and their duration. In many cases, the peaks will be between noon and 6 pm, but they can vary from plant to plant. Look for overlapping operations that could be sequenced, especially among heavy power-using systems. See if some work that goes on at peak demand can be rescheduled. A local utility has problems of its own during peak demand periods, and it will usually welcome efforts to equalize system loading.

Worker comfort is as varied as people themselves. The old a/c principle was to keep inside temperature 15 deg F cooler than outside. Today's practice seems to be to maintain 75-80F inside, no matter what the cost. LP should remember that any change will bring complaints.

B B BROWN, *Hagerstown, Md*

Anticipate coming changes in demand charges

Some present-day utility thinking may help LP come to decisions on his demand monitoring. Utility engineers say they are being shortchanged by consumers who have installed demand monitoring to meet the minimum requirements for a time, and then bring the full load on again. When several plants do this simultaneously, the utility is just as badly off as ever. The sliding-window concept has come in because of this, with the time period moving constantly, and thus requiring more complicated equipment. The utility charges for peak sustained demand during *any* 15-min or 30-min period.

This means LP should plan ahead several years and try to anticipate what the trends will be in utility demand-charge figuring. The rules can change, and LP may find that he is pursuing a moving target. Low-cost equipment to do the minimum task now can be obsolete in a short time, and this makes an engineer wonder if perhaps there is not a battle of wits going on between utilities and users. The utilities may perhaps want to hold on to the demand charges at a time of low load growth, especially where the total peak is not growing. On the other hand, users naturally seek to cut what they consider a penalty. If they can do it by good management, they want the saving.

R A WALKER, *Chicago, Ill*

Compromise saves part of production loss

We tried to hook up our process and building-services computer systems to give us demand control, but finally had to abandon the idea because of lack of flexibility. Despite all our planning, we kept running into situations where a cutout of power would cause serious waiting time, or would reflect in the next day's schedule of production. We settled on an analysis of what were the maximum loads we could drop with assurance that we would not impair our plant's output; since then we have operated with a simple cutout of certain loads, mostly heating and lighting. We still pay some peak-demand charges.

J G HAWTHORN, *Baltimore, Md*

Trade air speed for chilling power

Worker safety and comfort will not often be a crucial factor in planning demand monitoring. There is enough storage in the average building system to carry over several hours of reduced a/c effect. In some cases, especially in work areas with hot and dry conditions, an increase in the velocity of the air will make up for the loss of cooling, and at critical periods this may help cut load.

L E ENGELMANN, *Houston, Tex*

7.6 Clues to industrial-energy efficiency

Until recently, my engineering work was in design and manufacture of heavy ore-handling and -processing equipment. Then, last year, came an assignment in management of the company's energy facilities, which means steam-boiler plants at three locations, one of which generates its own power. Each plant has steam-piping systems, fuel-handling systems, waste-treatment plants, air-conditioning and air-moving systems, and many other elements. I am trying to learn all I can about these systems, but I can go only so far with theory. Not all the thermodynamics is useful, either, because of economics.

One of my shortcomings that I would like to get past is the lack of an overall viewpoint that would help me make a good, quick estimate of priorities for energy management in a given situation. We have meters and test gadgets available, but I think we would benefit from experience-developed rules of thumb on what to go after first. What shortcuts have POWER readers worked out in setting priorities for energy study in industrial plants? Are there any good clues, readily visible, to help rate the energy-management effort? DCV

This checklist has proved out

Here is a checklist for a quick survey. Any of the items can waste significant amounts of energy, and all of them have been found in actual plant inspections.

■ Are traps in good condition? Is there a regular inspection program?

■ For boiler efficiency, is exit air in correct range? Is stack-gas exit temperature in line with original specs?

■ Is steam venting to atmosphere for any reason? Do you have a really good measure of the amount? If not, chances are that you are wasting more steam than you think.

■ Do boilers have continuous rather than intermittent blowdown? Continuous blowdown can save energy.

■ If your turbine/generators are operating condensing do you know the cost relative to purchased power? Is it economical for you to generate condensing power?

■ Is your steam-piping insulation in good repair? Cost of adequate insulation can easily come back in savings from reduced heat loss.

■ And last, are deaerators and closed feedwater heaters at the correct temperature for feedwater at design conditions?

M H BORLAND, *Dayton, Ohio*

Try for material and energy balance

Highest priority is usually fuel/steam, followed by electricity, then water. In the boiler house, check to see that heat exchangers are recovering all waste heat from stacks, blow-down, and return condensate flash. Recover all condensate, and institute a routine to keep boiler efficiency at maximum.

Out in the plant, look for conservation through procedural and capital-project improvements. Identify specific departments using the most heat, and ask each supervisor to identify his five most wasteful procedures. Then, with help of engineering, look for ways to improve equipment efficiencies through heat recovery, utility recovery or recycle, and replacement of inefficient equipment. An energy audit helps.

Quickest way to reduce electrical costs is to turn off major equipment when unneeded. HVAC systems are a prime target here, because conditioned space is unnecessary on nights and weekends, and control timers are simple and reliable. Analyze production patterns and check major lighted areas. Demand control would be a second priority unless load factors are poor and an obvious pattern exists.

To cut water costs, gather up "good" water now going to the sewer and put it into boiler or cooling-tower feed, scrubber and direct condenser feed, and crude process makeup. Source of this water is often once-through cooling from heat exchangers in the plant and utility complex. Air compressors, distilling units, reactors, and crystallizers are examples.

All in all, the idea is to make a material and energy balance around specific areas, and then go where that leads. There are no shortcuts—it takes digging, and may be time-consuming.

J D Ehrhardt, *St. Louis, Mo*

Determine energy per production unit

One quick way for DCV to establish energy-management priorities and judge individual-plant performance would be to determine total energy per production unit for each manufacturing facility. We have found that millions of Btu per ton of raw material is the best parameter. DCV could make plant-to-plant comparisons and judge company performance against his industry's mean (which should be available from the National Technical Information Service).

A periodic report from a designated "energy coordinator" at each plant, stating raw-materials and finished-product quantities and allocating itemized energy sources to various plant processes, will help energy utilization to get the attention it deserves. A further step could be plant inspection to see that lighting and space-conditioning systems are maintained at a minimum consistent with comfort and safety, and also that standby fuels are not in use purely as a convenience.

R A Nelson, *Stamford, Conn*

Take advantage of shortcuts

DCV might set up these priorities:

■ Shut off utilities during nonproduction periods, and stop such obvious waste as free-running streams.

■ Repair all leaks as quickly as possible.

■ Reduce or eliminate "throwaway" streams, or their heat content in power and processing areas—hot exhausts, stack gases, hot-liquid emissions, and steam condensate.

■ Appoint an energy-management coordinator to set realistic energy-use standards for each operation.

■ Set up and report on power-consumption standards and goals for each operation.

DCV should keep some short cuts in mind, too. If fuel-burning-equipment stacks run much over 350F, economizers or air heaters can save about 1% of fuel for each cut of 40 deg F. Blowdown or liquid discharges at 120-150F can often preheat feedwater or process streams. Finally, as local cooling-water temperature decreases from probable summer highs, require at least 1 deg F more Δt across all coolers for each 2 deg F drop in supply. This will reduce pumped quantities and energy.

N P Baumann, *Corpus Christi, Tex*

Reduced boiler pressure may help

If a plant's major energy use is for space heating rather than process or production, there are several effective pointers to consider. Check of ventilating systems to see if intake of outside air can be restricted is one step. Timers for fans can help by shutting down systems when not needed. Systems to recover energy from air-exhaust systems can be investigated, too, along with heat pumps. Building insulation is another topic. Roofing and sidewall insulation is inexpensive and pays off in two to four years. Storm sash or insulating sash is another good bet, but may not pay out before five to eight years.

Operating steam pressure may be higher than actually needed. Try dropping steam-plant pressure by 5 psig to 10 psig each week, until problems begin to occur. Lower steam temperatures mean lower thermal losses through pipe and at valves and reducing stations. Watch steam-flow metering, however, because most detectors and transmitters are affected by operating pressure— recorders will show too high a flow. DVC could also try dropping system pressure for his compressed air; considerable energy savings are possible. Reduce pressure in small steps, until problems show up.

G C GONYEA, *Potsdam, NY*

You need your instrumentation

DCV should take another long look at his operating log books. They will tell him where and when most of the energy is being used. Well-kept log books will either have operating efficiencies recorded or will give data permitting their calculation. The basic idea is that the energy output/input ratio should be as close to unity as possible. Assume that the efficiency of every system and component can be bettered, and don't let any user or operator sway you from this philosophy.

Meters, recorders, and other good instrumentation, appropriately installed and properly maintained, are essential. Don't forget to take weather effects into account, either. With a thermograph, degree-days can be tracked, and energy consumption related to weather changes. Sometimes this accounts for unusual excursions in consumption rates.

Start first with improvements that carry little or no capital investment and that don't disrupt operations. Then alter or retrofit for higher efficiencies— after an economic evaluation has justified the work. Don't neglect distribution systems and terminal applications, either. Finally, develop operating standards applicable to your facilities and operations. Question any variance, but be ready to adjust the standards if necessary.

D PERKINS, *San Jose, Calif*

7.7 Best equipment for energy management

Our process plant is evaluating ways of trimming both fuel and power costs. We have been looking at annual steam usage of nearly 800-million lb and annual power consumption of over 56-million kWh, so that enough money is involved to make it worth the attention of management. Our steam needs are small in relation to electric load, and we have been advised not to generate our own power. Our boilers are package-type, watertube, burning oil to generate 250-psig steam. They are 12 years old, reasonably efficient and well maintained. There seems to be no way to improve their efficiency and make the investment pay off. Our main chance for cost-cutting, therefore, seems to be in savings under our existing conditions. We have tried cutbacks in lighting and space heating, and have tried to guess where the power and heat have been going.

The major handicap right now is the lack of good data. I think that improving the metering setup is our first step in establishing our minimum needs. We have several steam meters on major lines in and around the boiler house, and several department meters for electric load, but we need much more information. We also need good data on space-heating losses and processes, but we don't want to spend excessively on meters and test equipment. What has been Power readers' experience with selecting and installing meters and test equipment for energy management? Are there any quick ways to locate serious losses? How can we make better use of existing meters and test gear? —TNW

A portable meter setup is flexible

For spot metering of local usages, try a portable meter arrangement. Set a differential-head meter on a portable table, with means for leveling it at the operating sites. Carry orifice flanges and flow nozzles for each pipe size encountered in the field, and be sure that you have the necessary straight run of line ahead of and behind the flow meter. Remember, too, to run the metering apparatus long enough at each location to establish a full range of consumption rates.

This arrangement will give air-flow rates, if orifice plates are installed. TNW doesn't have to have the necessary engineering talent on his payroll, either, because most instrument sales engineers will be glad to help him set up his needs. A portable recording ammeter can be a source of basic information in finding what the electric load is. And don't forget the energy conservation consultants, either. Perhaps they are not as inexpensive as TNW's own plant efforts, but they are competent and can show many ways to reduce costs.

B B Brown, *Hagerstown, Md*

A boiler plant needs many meters

Simply providing meters for the purpose that TNW intends will not be sufficient—unless the operators are thorough in searching for the causes of malfunction and remedying them. Nevertheless, here are some variables that can be metered and recorded for a starting effort:

■ Oil flow vs steam flow. This should follow a known relationship. If oil flow goes up, look for trouble in oil nozzles.

■ Oxygen content in flue gases. This is important for overall efficiency and for avoidance of corrosion in air heaters. Low excess air means more tendency to produce acid deposits in air heaters.

■ Exhaust-gas temperature. Too high a value means more loss in flue gases. Again, corrosion danger will keep TNW from reducing it too far.

■ Feedwater temperature at economizer inlet. This will tell if the feed heaters are malfunctioning.

■ Makeup-water flow measurement. This gives a direct clue to losses of steam because of leakage, or venting and draining of the system.

■ Air-heater outlet gas temperature. Combined with check on excess pressure drop across the boiler passes, and check for sluggish control of final steam temperature, this will give a clue to sootblower performance.

For heat-exchange equipment, TNW needs several more good indicators. He could try these:

■ High terminal difference in temperatures. This indicates dirty tubes.

■ High rise in cooling-water temperature. Blocked tubes or low cooling-water flow can be the cause.

■ High difference between steam-exhaust temperature and condensate temperature, where steam exhaust is used for heating. This can point to a high pressure drop across tube banks. Perhaps air is blanketing the steam side.

These indicators also are useful with feedwater heaters in the boiler plant. TNW should watch for long-term contamination of heating surfaces and inefficient venting of heaters, too. Trend graphs can be kept to detect these effects.

R K JAIN, *New Delhi, India*

Simple instruments are best for start

A quick and dirty calculation shows that TNW's plant produces steam at a rate of 400,000 lb/hr, and uses electricity at a rate of 28,000 kW, so this is no small operation. Lack of good data seems to indicate absence of a good plant-maintenance program—work is done on an "emergency" basis, with production being the governing factor. This is not unusual, but TNW is thinking ahead to correct this. The best "meters" are in the heads and hands of the maintenance men—eyes, ears, touch, even taste and smell. Eyes can spot steam and piping leaks, and inoperative steam traps. Ears can detect steam and air leaks, and recognize offbeat sounds in equipment.

TNW could start with his boilers. Is the blowdown orifice too large? Are blowdown valves tight? Are the safety valves chattering? The distribution system for water, steam, air and process fluids should be tight, too. Steam traps should be operating; the touch of a hand can tell. Are there liquid drips in sumps, or in buildings and tunnels? The list is endless. A few instruments, inspected and calibrated on a rigid schedule, are helpful. A precision pressure gage against which the shop gages can be checked is a useful instrument, and a small portable anemometer to check air flows in ducts is another.

H M NEUHAUS, *Santa Monica, Calif*

Steam pressure check can help

TNW might remember that improving his meter setup will not immediately give big savings. He will get additional data to file away, but the departmental electrical and steam loads will still be the same after the data have been collected. Most savings are made by starting with cost-cutting ideas. One difficulty in putting such ideas into action is that the powerplant engineer, maintenance supervisor, and plant engineer do not exercise managerial control except in their own departments.

For instruments and meters, TNW might try a small, portable 24-hr recording temperature gage to check heated spaces. If excessive heat is found to be supplied, then a single thermostat to control solenoid valves or electric space heaters could be an energy-saving step.

Additional steam-pressure gages put in to check low-pressure and high-pressure steam lines for excessive pressure drop can lead to more-efficient operation if the excessive drop is corrected, and this should reduce overall steam use.

G H HILL, *Sunbury, Penn*

Get more from your existing meters

Make periodic calculations relating energy consumption to production and space variables—along with the weather bureau's degree-days, if applicable. Ratios like kWh per lb of product, kWh per ft^2 of space, or lbs of steam per degree-day can help. A monthly abstract will give true average values. Indications on where to locate additional meters can come from this, too.

J A BARRETT, *Houston, Tex*

7.8 Save fuel with lower steam pressure?

The energy squeeze we are in now is forcing us to scrutinize our boiler and steam-distribution system to find savings. Our plant, a large industrial, has about 2-million ft^2 of floor space, and we generate steam at a rate of 1.3-million lb/day, in watertube package boilers. We burn natural gas now but may have to switch to oil soon. Our flue-gas temperature is 450F. The steam, 125-psig saturated, serves for both heating and processing—in degreasing tanks and similar uses. Out in the plant, the 125-psig steam is reduced in pressure at several stations, so we can maintain a constant lower pressure on equipment and keep temperatures nearly constant.

A suggestion has come to us that, by reducing steam-generation pressure from 125 psig to 110 psig or even 100 psig, we might save some fuel. I do not see how this is possible, but I would like to get opinions, especially from engineers who have tried it. Do POWER readers know whether we can save fuel this way, perhaps with added minor changes in equipment or operation? Also, how should we plan our pressures out in the plant for better economy?—KNN

Yes, because flue gas would be cooler

Yes—in at least two ways. First, no mention is made of superheat, so if steam is raised at 100 psig instead of 125 psig, the steam temperature would be lowered about 16 deg. Flue-gas temperature should be reduced by a like amount, giving savings of about 0.4%. Second, the difference in heat content of saturated steam at 100 psig compared with 125 psig is about 3.6 Btu/lb of steam, or about 0.3%.Since the steam is reduced by regulators, this represents savings of steam by pressure control at the boiler rather than dissipation of part of the original higher pressure in a regulator. These savings are additive, with a total of 0.7% saving in output, or about 0.875% saving in fuel (at 80% boiler efficiency).

E R GARDNER, *Cambridge, Ohio*

Yes, and we proved it!

Our industrial plant has 1.7-million ft^2 of floor space and is steam-heated by three 55,000-lb/hr watertube, gas-fired package boilers. Some steam goes to processes such as steam-cleaning, and some to cafeterias, hot-water heaters, and cold-absorption units. Boiler load varies from 100,000 lb/hr in winter to 2500 lb/hr in summer.

Fuel can be saved by going to lower boiler pressure. In our case, we went from 150 psig (366F) to 100 psig (338F). Efficiency of the boilers went up slightly because the water is cooler. Besides, radiated heat loss from steam mains is less. Steam leaks at flanges and packing glands will be reduced, and less boiler-feed-pump energy is necessary because of the reduction in pressure. To determine energy savings, our boilers operated for one week at each pressure when the steam load was nearly identical. Gas readings were taken directly from the utility meters. The boiler load was approximately 7500 lb/hr. At 150 psig, the boilers used 1,299,500 ft^3. At 100 psig, gas was 1,155,800 ft^3, for a saving of 143,700 ft^3, or a 147-million Btu saving per week.

The saving from improved boiler efficiency and feed-pump need is about proportional to steam flow. Saving from radiated gas loss and steam leaks is relatively independent of flow. Any estimate of total yearly saving would require comparison at several different demand situations. For energy conservation, boilers should operate at minimum steam pressure needed. And remember to rework orifice-plate meters or apply a correction factor if pressures are cut from design conditions.

M PETZOLD, *Grand Rapids, Mich*

Perhaps, but look into other factors

It is possible to reduce total steam consumption by reducing the pressure of generated steam, but other factors must be considered. At a rate of 1.3-million lb/day of steam, the maximum savings in fuel through reduction in steam pressure from 125 psig to 100 psig would be 3¼-million Btu/day (based on 212F feedwater temperature). This is equivalent in firing rate to 4166 ft³/day of natural gas or 29 gal/day of oil, and is only a saving of 0.25% of the total fuel.

The plant piping, reducing stations, and processing equipment have all been sized for a particular steam operating pressure and temperature. Therefore, if steam pressure is reduced, adequate supply may not get to various points in the plant. It may be possible to decrease operating pressure to 100 psig without detriment, but I would not recommend going below this.

Oil firing of watertube boilers is more efficient than gas firing, but the higher cost of fuel oil per heat unit will make oil firing more expensive. The best approach to reducing steam-generation costs is a complete plant survey. Fireside and waterside of boilers must be kept clean, and burners properly adjusted for minimum excess air. Steam-supply piping needs inspection. Insulation and leaks are important. Steam traps should be operating and not frozen open. All low-pressure steam and condensate from heating and process should be returned to a hotwell or deaerator. This is the most common area of loss, because condensate from heating and process steam is generally dumped to drain after use.

The reducing stations in KNN's plant are the most practical and least expensive method of controlling process steam. No heat is lost in the expansion process, because the low-pressure steam merely becomes superheated.

L L Yost, *Winfield, Kan*

We saved—and cut line losses, too

A saving is possible through reduction in primary steam pressure. We did it two years ago with good results, and we are convinced that it is worthwhile. There must be sufficient primary pressure to deliver the steam to the point of use, and the steam must have the necessary energy content to do the required work. Beyond that, more details of KNN's plant would be needed.

Saving on heat transmission through pipeline insulation can also be found. Conventional equations from standard engineering textbooks or handbooks will help. In this case, the 15-deg difference between 125-psig and 100-psig steam is equivalent to 7 Btu/hr/ft² of pipe insulation area. At 1075 Btu/ft³ for gas heat value, the potential saving is 168 ft³/day of gas for every 1000 ft² of pipe-insulation area.

B B Brown, *Hagerstown, Md*

How can anyone tell what savings are?

KNN's load is heating and processing, and since I suspect that nearly all the processing is really heating, too, I will assume that all the steam going out to the plant is for heating. Because the steam is being used for this, its pressure has no effect, within wide limits, on the total Btus required. Therefore, for a first approximation, KNN cannot save any fuel by changing pressure from 125 psig to 100 psig.

There are some minor sidelights, however. One involves traps. If KNN's *high-pressure* traps are discharging directly to air, he can save a little by going to the lower pressure, because more Btus will be extracted from each pound of steam, and the condensate will have less energy to be wasted. The ratio is about 881/868, for a 1.5% saving. If the condensate goes back to the boilers or heats water directly, then even this saving is impossible. *Low-pressure* traps will still discharge at their old reduced pressures, of course, so no saving is possible there unless the condensate is now being discarded.

Another sidelight concerns the boilers themselves. I assume that they have no heat-recovery equipment, and that the difference between saturated-steam temperature and flue-gas temperature will be the same over a range of steam pressures. By operating at a lower pressure, with saturated-steam temperature lowered from 353F to 338F, the boiler will extract more heat from the fuel. Outlet-gas temperature will go to about 435F, for a gain of perhaps 0.8%.

These savings are theoretical, problematic, and minor. I doubt that most plants could detect them with high confidence, but if KNN wants to go after them, he is welcome. He should remember that he will be risking more process troubles caused by variation in the reduced pressures out in the plant. This is a real danger. Keep in mind that one reason for pressure reduction is to compensate for variations in supply pressure. A high basic supply pressure makes excessive dips less likely and therefore gives the pressure regulators a better chance for uniform output. I think that KNN might better turn attention to (1) return of his condensate, or use of it for direct heating, and (2) study of flue-gas heat recovery to extract some energy from the 450F flue gas.

G HAMMEL, *Denver, Colo*

Enthalpy is key to the problem

A saving should be possible because of the 3-Btu/lb difference in enthalpy between saturated steam at 125 psig and 100 psig. Further reduction will save more fuel. He should avoid unnecessarily long pipelines, too, in which high pressure drops will decrease the steam's enthalpy.

A CHAUDHURI, *New York, NY*

Try an economizer instead of dropping pressure

Theoretically, there is a fuel saving by dropping steam pressure, although the saving is less than 0.3% from a change to 110 psig. Percentage gain by pressure reduction is small because the major effect in a plant of this type comes from the latent heat of vaporization. Better savings are possible in other ways. If the boiler plant is centrally located, steam probably arrives at the point of use at 90% quality. There is no superheat to take up losses on distribution lines. Adding 20-deg superheat would ensure dry steam at point of use.

Recovering and reusing process/heating condensate and putting better insulation on steam-distribution lines would help, too. An economizer in the boiler uptakes to reheat boiler injection water and reduce stack temperature from 450F to 375F is an additional measure.

H M NEUHAUS, *Santa Monica, Calif*

Turbines are a better prospect

KNN should not reduce steam pressure to get fuel savings. Based on the difference in enthalpy between steam at 125 psig and steam at 100 psig, generating an average of 1.3-million lb/day of steam at 100 psig would save about 4.3-million Btu/day. At a fuel cost as high as $5/million Btu, the daily saving would be only $22.

Instead of reducing steam pressure, KNN should review his plant to see if small turbine drives can be substituted for electric motors. With 125-psig throttle pressure and exhaust to KNN's low-pressure system, mechanical energy would be recovered to save over equivalent purchased power. A study should be made, too, to find the minimum pressure and temperature for process needs, because potential mechanical-energy recovery by turbines increases with lower exhaust pressure.

W R ZIEMANN, *Newark, Del*

8 Plant Systems and Equipment

8.1 Mothballing a new steam plant

Our brand-new steam plant seems to be a victim of the energy crisis, the credit squeeze, and regulatory uncertainty. Management has decided to lay up the facility for at least a year. We had just gone through startup of our two 125,000-lb/hr boilers, with a short run at full capacity on all components. Now we have to close everything down for an indefinite period. Because this is something with which I am not familiar, I am looking for guidance.

Our boilers are stoker-fired, for process steam at 300 psig. We have a lime softener system, filters, deaerator and storage tank, air compressors, and the usual station pumps and service-water system. My instructions are to shut down with as low running expense as possible. We will need some winter heat in our cold climate.

What have POWER readers experienced in this type of shutdown? Can we rely on the advice of boiler and other equipment manufacturers? Is our steam-distribution system vulnerable? And should we rent a heating boiler instead of running a main boiler?—JDR

Lime for drums is important after draining

In my experience as a boiler inspector, I have seen hundreds of boilers ruined by improper layup. The best, easiest, and cheapest layup method is:

■ Open and drain all water lines that run to the boiler. Then open all manholes and handholes in the boiler, and let it dry out for several days.

■ Insert lime in each drum, and close up all openings to the water side.

■ Put an electric or steam heater in the combustion chamber to keep the boiler above dewpoint. JDR will probably need some building heat to protect other fire and water lines, so the purchase or rental of a small 15-psig steam-heating boiler is a good idea. An old section of wall-mounted cast-iron radiator makes an excellent heating unit for the combustion chamber.

I recently inspected a boiler that had been laid up 12 years ago with this method and the boiler was still perfect. In case of a long layup, change the lime and fire the boiler a few hours once a year. Don't forget to drain all pumps, lines, and auxiliaries to prevent freezing.

M A SHELTON, *Portland, Ore*

Fill steamside equipment with N₂

JDR's problem is unique in my experience. I have never heard of a new plant being shut down so quickly. I have had some experience in mothballing Navy ships, however, and this could help JDR. To be able to get the plant back on line at short notice, here is what I would do:

■ Rent a boiler for steam heat in winter.

■ Drain water from all steamside equipment and fill with N_2. (Lines needed in winter are an exception to this.)

■ Clean and drain the water-treatment plant, leaving it so dry air can be circulated through the equipment.

■ Keep all electric equipment dry, and megger it every quarter.

■ Lay up the boilers according to the manufacturer's recommendations.

■ Keep up the battery systems, and check them monthly.

■ Finally—be sure to rotate pumps and other rotating equipment at least a quarter turn each month. T W HAMMOND, *Santa Nella, Calif*

Dry method using desiccant counters freezing

There are several shipboard methods for laying up boilers: wet, dry, and coating with metal protectors similar to cosmoline. Specific manufacturer recommendations are advised. I suggest the dry method, because it needs less attention after layup and has less danger of freezing. Here are the steps:

First, operate the sootblowers to clean the boiler's fireside. When the pressure side has cooled, drain it completely. Next, open steam and water drums and waterwall headers. Then open the superheater drains. Circulate air from a portable blower through the drums, tubes, and superheater until they are dry. Put desiccant in the drums to absorb moisture. And check the desiccant fairly often, replacing it as needed. With the wet method, maintain an alkaline condition, with deaerated water, and keep a slight pressure on the boiler. Finally, remember that all this work can be burdensome during the shutdown.

F H CUMMINGS III, *Rockland, Me*

This is a special case for boilers!

As long as steam is kept out of the piping, JDR will have little trouble from rusting. Blank off connections between heating lines and other piping, and make sure unused piping is reasonably well sealed up to prevent breathing. Blanking off large pumps after draining them is also advisable, because often water forming elsewhere will drain into these pumps and create bearing problems. Be sure to keep a record of where the blind flanges are, to avoid embarrassing incidents on restart.

A low-pressure rented heating boiler might save on operator costs, but there are other considerations that need advance planning. For example, if the heating system is designed to operate at 60-100 psig, the reducing valves and traps might not take too kindly to the low pressure. Then, too, if the plant has a turbine-driven fire pump, JDR will be sacrificing some protection by foregoing his alternative fire-pump drive.

The boilers themselves could be the most endangered equipment during a long shutdown. The boilout and short test run probably cleaned out whatever metal protection the boiler internals had, and the run was doubtless too short to build up a good oxide coating. Now, instead of relying on his regular water treatment to develop oxide, JDR will have to keep the interior surfaces completely dry by circulating warm dry air. Desiccants are not the entire answer, especially for low-lying spots. When asking the boiler manufacturer for advice, JDR should be sure to give the exact present status of the boilers. Don't rely solely on general suggestions for layup of long-operating boilers.

K M Battagliani, *Columbus, Ohio*

Heating might call for main boiler

If JDR has sprinkler systems in his plant buildings, he will need considerable steam for freeze protection. I tend to think the amount of steam in question would justify running one of the boilers, with reduced-pressure steam going to the heating system in winter. This means licensed operators, of course, but the cost could be less than repairs to freezing damage. Load would be light, especially if most of the condensate returns. Another advantage of keeping a boiler on line is that the plant will be able to restart faster, with no time lost in attending to unpleasant surprises—such as theft of valves and piping, which has happened to us during shutdowns.   R S Watkins, *Chicago, Ill*

Warm layup is wise here instead of a cold one

Moisture from condensation and residual water in the various systems—these are the biggest concerns for a powerplant layup. The management here is apparently willing to have a warm layup instead of a cold one. This is wise, because cold mothballing would cost more for a comparatively short time.

Advice of boiler and other equipment manufacturers will be reliable in general. Their equipment sees a multitude of field conditions. Although covering all the alternatives would require extensive directions, JDR can rely on what reputable manufacturers tell him.

For piping systems, complete drainage is necessary, with particular attention to drip legs, strainers, and traps. I have found that steam and feedwater lines have enough residual heat so that, if they are bled and drained quickly from hot service, they will dry reasonably well. We have run desiccated or well-dried warm air through these lines at a low rate to prevent moisture build-up from condensation. This procedure will cut the vulnerability of the steam-distribution system. If service-water piping is galvanized, simple draining will be good enough. Draining and cleaning the softeners and filters to prevent rusting and later iron contamination is a good idea. The layup of the deaerator and storage tank can be the same as that for steam lines. A few bags of desiccant inside will help.

Open the compressors, and either swab the running surfaces with oil or spray them with rust-preventive. This treatment is good for pumps, too, if they are likely to rust or corrode. Plunger pumps need protection on sliding surfaces. Remove all boiler manhole and handhole plates, and then coat all sealing or mating surfaces with heavy lube. Connect a unit heater in the combustion chamber, and blow warm air constantly through the interior, keeping the temperature above dewpoint. Keep dampers closed to maintain temperature.

I agree with the idea of renting a small boiler, for operation below 15 psig, to keep the plant on warm layup. Automatic firing for the boiler will eliminate some labor expense. For a prolonged shutdown, however, economics calls for a cold mothballing. This will cost more at the start, but will save money in the long run.

B B Brown, *Hagerstown, Md*

8.2 How much more life do oils have?

One group of my company's operations consists of several small plants, far apart and mostly remote from large cities. All the plants have hydraulic systems on tools and equipment, and all require considerable amounts of lubricating oil. We have studied the plants' use of these fluids and lubricants, and think we could cut costs by stretching out times between changes, and by reuse.

One obstacle, however, is the need to test the oils. When we discuss the problem with lubricant suppliers, they tell us we need all kinds of tests—for viscosity, neutralization number, water content, flash point, etc—to determine suitability of a lubricant for continued use. Our setup simply does not lend itself to constant inplant testing of this type. In addition, experience tells me that sending samples to distant labs will not work either.

Do Power readers have experience with simple criteria for decisions on whether a hydraulic fluid or oil has additional useful life? What would be a good program of simple steps to bring fluids and oils back to serviceability, without expensive out-of-plant processing? What changes would stretch fluid life?—WHS

Test kit reveals particles

Our company's operations include three generating stations and six dams, spread over nearly 150 miles. We buy our hydraulic fluid from a supplier who provides analytic services and gives us quick and accurate reports. WHS can stretch life without buying expensive equipment if he will get a small patch-test kit, with a standard Millipore filter, to show particulate contamination in hydrocarbon-base fluids. Test results are repeatable, and sensitive enough to detect significant change in cleanliness. These tests, plus adequate filtration, should extend the service life of the hydraulic fluid.

D GUENTHNER, *Marble Falls, Tex*

Eye and nose detect key changes

Oil never wears out, it merely gets dirty. Contaminated oil can change viscosity, neutralization number, water content, flash point, etc, in a reservoir. Oil contamination occurs when new oil goes in, but other causes are poorly designed or installed air cleaners, lack of protection for cylinder-rod seals, and cheaply designed filter systems.

Simple in-plant tests include visual, with comparison with new oil, and odor tests, in which a burned smell indicates loss of lubricating value. A patch test is good, too. Compare color and color density with qualities of new oil, and look for large dirt specks. Pouring the oil through a coffee filter is helpful here, but test kits are on the market, too. I don't know of any filtration system that can remove carburization products.

WHS should consider oil reclamation. Heavy contamination and water settle out slowly, accelerated by heating to 130F in a sloped-bottom tank. The top oil needs crude filtration, dehydration, and then some polishing filtration, followed by introduction of additives. All in all, the most cost-effective method of stretching oil life is filter-system improvement, correction of sources of contaminant entry, and, finally, oil reclamation.

J R SULLIVAN, *Winchester, Va*

Analysis of lube oils is still indispensable

WHS probably has several oils, such as lubrication oils and hydraulic oils of various viscosities, along with "hydraulic-lubrication" oils operating in various temperature ranges, under severe or normal conditions. The last-mentioned oils probably have antiwear additives and the normal rust, antifoam, and antioxidation inhibitors. The lubrication oils may be exposed to air and periodic water ingress through leaks in coils, seals, or heat exchangers. This adds up to an appreciable number of different oils under different conditions.

Reconciling this with "simple criteria" is difficult. There is no simple test to monitor overall condition of a single oil-fill in a given service, nor for several oils in varied services. Routine testing is nevertheless necessary. WHS mentions "constant testing," but in my experience, initial tests rerun every six months are enough for "hydraulic-lubrication" and lubrication oils. An oil fill usually establishes its behavior in a year, allowing reduction of tests to once a year.

Take a one-quart test sample from each fill. Viscosity, neutralization number, etc, normally stay stable. Some governor lube oils, steam-turbine oils, and fluid-coupling oils are still running after 20 years. Drastic changes in viscosity and neutralization number are a key, but I have encountered drastic changes on only three oil fills over 20 years of routine oil analyses.

Dielectric strength is a "go/no-go" test for an insulating oil. It will not reveal the whole story, but if a transformer-oil fill breaks down at 5 kV, then the transformer should come out of service immediately. Drastic rise in neutralization number is another cause for concern in "hydraulic-lubrication" and lubrication oils. A rise to the upper limit is cause for immediate oil change. Drop in interfacial tension is also immediate cause for concern.

In Canada, oil companies analyze customers' oil fills free, at a normal frequency of once a year. Analyses include those for wear metals, viscosity, neutralization number, flash and fire points, and corrosion-strip tests. No doubt some US oil companies do the same. This free service is the answer to WHS's problem although he should be wary of any "all-in-one" test for monitoring plant oil performance. There is no such test.

Oil analysis is a must. Strictly scientific service can be expected, too. If WHS thinks a supplier will condemn a healthy fill for the sake of a sale, he can reassure himself by getting condemning limits from several suppliers, and a comparison of the limits should demonstrate that he can expect good service. For bringing oils back to serviceability, a filter press can be useful, provided the oils contain only dirt, dust, lint, water, sand, paint chips, or wood fibers. Chemical contamination requires more sophisticated filtration, such as fuller's earth followed by re-addition of inhibitors.

Timely analysis will allow correction before an oil turns to sludge or gets loaded with crud. In every case, however, the old oil fill needs analysis after treatment, to assure that it meets new-oil specs. Ask suppliers about inhibitors for their products.

E GILLESPIE, *Winnipeg, Man, Canada*

Light box helps visual test

We recommend a simple visual-inspection program to our customers. It gives first-hand knowledge of the general condition of the oil, and prevents minor

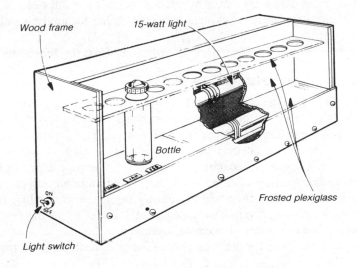

Wood frame · 15-watt light · Bottle · Frosted plexiglass · Light switch

8.2

adverse conditions from becoming major. Without such a program, a plant must depend on a lab for costly and time-consuming analysis.

A combination light box and storage rack (sketch) is a central item in this program; it allows storage and visual examination and comparison of samples. Good illumination is important to allow an observer to see through the oil, but don't store the oil samples in direct sunlight, because UV rays can change oil color. Although standard infant nursing bottles are good containers—cheap, sterile, and capped—any container with similar characteristics is acceptable.

First step in the program is getting master samples against which to compare later ones. Draw the masters when new oil comes for installation. Clean the sample cock in the supply header with a filtered solvent and lint-free wipers before drawing any sample. The system should be operating during sampling. Flush a gallon or two of oil from the sample cock before filling the sample bottle. Take samples at monthly intervals, and set the sample bottles in the rack for comparison against previous and master samples.

Here is what to check in the visual inspection:

■ **Color.** Although entrained water will settle readily, an emulsion in formation will make the oil hazy, from very light to milky. Entrained air tends to create haziness, too, but look closely and you can distinguish this cause. Lightweight particles, too, can stay suspended and add to color.

■ **Water and/or emulsion.** Most water will collect at the bottom within an hour. Maximum allowable water content is 5%. Read the scale on the bottle to find water content. A centrifuge can remove water, if necessary.

■ **Odor.** A rotten-egg smell from the bottle shows overheating, which will also change the color to a dark brown.

■ **Contamination.** The naked eye can detect 25-micron particles of dirt, so look closely at the bottle bottom. If you see specks, check the previous months' samples to see if quantity has increased. A few pepper dots are no cause for alarm, but if the entire bottom is covered, check all the filter elements immediately, and draw another sample to send to a lab for analysis and identification of the contaminants.

D J Chaleki, *Worcester, Mass*

Statistics prove value of checks

Hydraulic oil is one type of lubricant that some plants require in large quantities. It lends itself to uniform plantwide purchasing and usage procedures, therefore, and high efficiency in the procedures brings economic and conservation benefits. Reuse of oil cuts consumption. Reclaiming, re-refining, and recycling are possibilities. Recycling is simplest, and we have done this for the past seven years. Getting maximum life out of hydraulic oil reduces oil and labor costs. At the same time, the annual testing for serviceability and our portable filtering procedures have raised the quality of oil in the reservoirs. Maintenance and repair costs have gone down, too.

Purchase of the oil is the starting point. Choice of a single, good rust- and oxidation-resistant oil of medium viscosity, with purchase in bulk, gives best price. The large sales volume makes the supplier willing to offer field lubrication-engineering services, too.

Recycling to reduce consumption is second step. An annual survey of all hydraulic machines determines oil serviceability. Draw a 10-cc sample from each reservoir by squeeze bulb and tubing, and put the sample into a test tube. Draw the sample from the middle of the liquid volume, with the machine running. Store the test tubes in a numbered, compartmented box with a cover. Clear the squeeze-bulb tubing of most of the oil after each withdrawal, to cut carryover to the next sample. A few remaining drops will not be a problem.

Incidental observation of the sample can detect cloudiness or haziness, meaning water or air in the oil. Water can mean a leaky heat exchanger, but air indicates a faulty pump. Darkening of the oil could mean oxidation. A cap or suction screen could be missing, or oil may be low.

Deliver samples and record sheets, with sample and machine numbers and observations about oil and machine condition, to the oil supplier's engineering department. There the analysis includes:

■ Checking color to determine amount of oxidation.

■ Checking viscosity with a viscosity gage. A 10% increase (approximately) accompanied by darkening indicates oxidation. A decrease may mean additive deterioration.

■ Performing a water-drop test to find water percentage. If bubbling occurs, there is 0.1% or more water in the oil, and the rust preventive is no longer effective.

■ Straining the oil through a 5-micron Millipore filter pad and examining the pad under a microscope, to determine dirt quantity and particle size. Pictorial charts help here. A guideline is 150 mg/liter as the critical threshold.

Hydraulic-fluid field-test kits are sold by at least one oil company to plants that want to do these tests themselves. Drain all machines that contain oil that is oxidized, water-contaminated, or dirty. Clean their reservoirs and suction screens, and change filters. Borderline dirt-contaminated oil can be cleaned with a 10-micron portable filtering unit and put back in.

Badly oxidized oil will leave a varnish layer on machine internal parts. Clean the layer off by adding 5-10% solvent or straight mineral oil to the system and run from 8 hr to several days, depending on the severity of the oxidation. Then drain and refill, as indicated above. Statistical analysis of 120 hydraulic systems operating at about 2500 psig in one area of our plant is shown in the table. A is % of machines with "unsatisfactory" oil, needing change; B is % with "satisfactory" oil (including borderline cases), not needing change; C is % with "satisfactory" oil (not including borderline cases), needing no attention to oil; and D is % with "borderline" oil, needing filtration.

The table shows that 95.7% or 804 of the 840 potential annual change-outs were not needed. Only 36 reservoirs were changed during all seven years. Of the 95.7%, 3.6% were borderline and were filtered, making the oil usable for at

Statistical analysis of 120 hydraulic systems

Category	1972	1973	Year of analysis, % 1974	1975	1976	1977	1978	7-yr avg, %
A	5.0	2.5	5.8	5.0	5.8	3.3	2.5	4.3
B	95.0	97.5	94.2	95.0	94.2	96.7	97.5	95.7
C	95.0	97.5	90.8	86.7	90.0	88.3	96.7	92.1
D	0.0	0.0	3.3	8.3	4.2	8.3	0.8	3.6

least another year. Other statistics show that 75% of the machines never needed oil change during the seven years. Makeup of less than 1½ times reservoir capacity per year is considered desirable.   C MUIR, *Warren, Ohio*

8.3 How to cut use of tracing steam

Our chemical plant has several processes that require pipe transport of liquids and slurries at elevated temperatures, and because there is danger of excess thickening or congealment if temperatures in the lines fall below minimum values, we have many steam-traced lines. We had never realized how much steam was being consumed in these lines until recent metering checks inside departments gave us some clues. Currently, we think about 18% of our steam consumption goes for tracing.

We trap this steam and return it to the boiler plant, saving part of the heat. In addition, we want to stop unnecessary use of tracing steam. We find very little information on what steps to take in this area, however. The economics of grouping or rerouting piping, increasing insulation thickness, etc, are not well-defined, either. Do POWER readers have experience with revamping steam-tracing systems to improve energy economics? Must we balance reliability against low operating cost? And will increased maintenance be a price to pay for steam savings?—VM

Trap survey is a good idea

For the short term, VM should consider these helpful hints:

■ Repair all steam leaks. A ¹⁄₁₆-in. hole on 100-psig steam can waste roughly $550/yr if steam is $2.50/million Btu.

■ Make a thorough trap survey on the tracing system. Tracing systems use disc (impulse) traps, which are prone to wear and loss of steam. Inverted-bucket traps are a replacement possibility.

■ Repair broken or loose insulation, and also moisture and weather barriers.

And for long-term planning:

■ Remember that steam tracing is inherently wasteful, consuming steam unnecessarily most of the time. Consider an electric tracing system. Note, however, that modulating steam-tracing control systems have recently become available.

■ Consider rerouting hot waste streams to use their heat.

■ Any economic-insulation thickness determined a few years ago may have to be increased now.

G C Shah, *Channelview, Tex*

Two ways to reduce steam-tracing costs

I have come across two methods that help cut steam-tracing costs (sketch, 8.3A). Both include self-contained temperature-control valves that don't

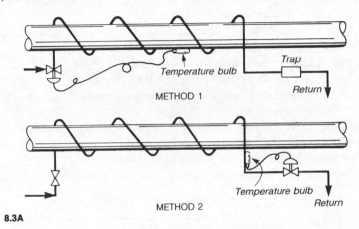

8.3A

require an air or electric hookup. In the first method, a self-contained temperature-control valve measures either the pipe-surface temperature or the fluid temperature. The valve regulates steam flow to hold the fluid temperature at a set point. Valve setpoint can be varied over a range. The valve is ordinarily at the steam-tracing supply manifold, eliminating the need for a separate valve for each tracing line. If slurries are to be heated, try measuring the temperature of the pipe's outer surface, because slurries are usually erosive.

In the second method, locate the control valve in the condensate line and measure condensate temperature. The control valve should be adjusted to maintain a constant condensate temperature which is near the required fluid temperature. Note, however, that this method may give a high steam-tracing pressure drop and affect the ability to return condensate—unless there is a condensate pump near. Either method will reduce the steam consumption, because the fluid is not overheated. In addition, as the weather changes, the steam flow rate will adjust to the changing heat losses.

M B Moskowitz, *Houston, Tex*

Review waste-heat sources first

Although some sort of heating is often desirable, it is true that too high a temperature can damage fluid, corrode pipe, and squander energy. VM should first consider the possibility of waste heat. Can a stream of return condensate from storage-tank heating do the job? Does piping that carries other warm process material pass nearby? Can the piping be rerouted through a warm room? If none of these alternatives are good prospects, then a source of primary heat, either steam or electricity, is a must.

A caution on routing piping through warm rooms, however: Leakage can occur. Piping carrying chemicals is often a high-maintenance item. Grouping the piping carrying the high-freezing-point material with other process piping that carries warm material can be attractive, but may have high insulation costs, and be messy if a leak occurs.

A stream of warm condensate is an excellent way to go. Lack of control is usually the main problem here, however. VM would probably prefer to stay with steam heating if he must provide primary heat. Temperature control is a simple matter. Reasonably priced valves are available that are actuated by a fluid temperature, ambient temperature, pipeline-surface temperature, or tracer-condensate temperature.

The most-expensive heating source is usually electricity. Resistance tapes with built-in thermostatic control are available. In rare cases, the steel piping can serve as a resistance element. Often desirable is a standoff of the tracer line from the line carrying process material. This avoids localized overheating

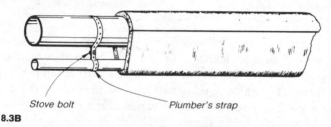

Stove bolt   Plumber's strap

8.3B

that might accelerate corrosion. (sketch, 8.3B). On completion, insulate the assembly neatly. Separate the lagging in half lengthwise, and put half on the process-fluid pipe. Then put a split length of lagging of the next larger size around the tracer and pipe bottom.

F W HOFFMAN, *Puyallup, Wash*

Improve heat transfer to piping

Revamping steam-tracing systems to reduce costs is often expensive in itself. If VM's tracing consumption is now 18% of his steam output, he could scarcely reduce that to less than 12-14%, even with costly reconstruction. On the other hand, upgrading the effectiveness of existing systems is a good idea. Application of a heat-transfer medium, for example, will improve heat conduction from tracer to process line. The medium, a mixture of sodium silicate and carbon, has a coefficient of about 50, compared to only 0.02 for air at 300F. Readily trowelable, the medium will upgrade tracing efficiency and cut steam consumption.

B B BROWN, *Hagerstown, Md*

Steam or electric, but control it

It appears that VM's heating steam is flowing uncontrolled at whatever rate it will condense in the system. An uncontrolled system sized to maintain proper temperature during cold weather will put excess heat into the liquid in warm weather.

The obvious solution is to apply heat at the minimum rate needed to satisfy needs—prevention of thickening or congealment. Weather conditions and line velocity will influence the amount of heat needed. Tracing can be electric or steam, but should be applied over a clean, rustfree pipe surface. A highly conductive heat-transfer cement is advisable to assure continuous bonding between tracer and pipe.

Stainless steel bands or straps will hold the tracer in place. Steam-admission point should be at the highest point of the line, and the tracer tube should slope from there, along the sides of the pipe, to the trap. Don't form pockets in the tubing except at the drain outlet to the trap. This means no complete circles of tracer tubing around the pipe.

Insulation covering the traced pipe must be oversize on the inside diameter to give space for tracer tubing. Piping 3-in. and smaller usually takes 1-in. insulation. Pipe over 4-in. nominal size takes 1½-in. insulation. Temperature controls need an adjustable temperature range, say 55-175F. The contact can be set to close at 100F and to open at 120F, for example.

N MOGENSEN, *Jacksonville, Fla*

8.4 How to stop condenser vibration

The main barometric condenser of a vacuum-refrigeration unit in our plant (sketch, 8.4A) has had serious vibration troubles, which evidently

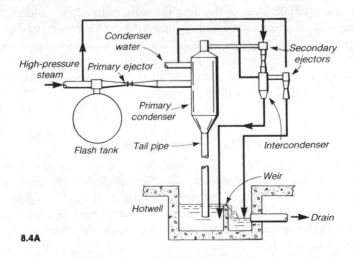

8.4A

originate in the 54-in., outside-diameter, fiberglass tail leg. The vibration has loosened bricks in the building walls and has set a main transverse I-beam into twisting motion. The vibration has increased as salt-water flow down the tail leg builds from a minimum of 5000 gpm to 20,000 gpm. We tried stiffening the guides at the base of the tail leg and at a point somewhat above the pipe center point, but this did not reduce vibration.

We then added another heavy lateral stiffener a few feet above the lower end. This stopped motion of the tail leg at the weir, but wall vibration and beam twisting continued—even more violently than before. We have limited load on the unit to 50% until a solution is found. Do POWER readers know of any effective way to solve our problem? Should we look inside the unit or should we seek the source in the structural setup?—BD

Orifice worked well to slash vibrations

We had a similar problem on a barometric condenser, and we found that merely stiffening the tail leg did not work. We stiffened the tail leg at the bottom and partway up. Vibration continued, so we began an investigation of the water inside the tail leg. When the unit is in operation, the main barometric condenser is under a vacuum of 1 in. Hg abs. Level of salt water in the tail leg is 32 ft above weir water level.

We found that the excitation of the tail leg is caused by entrapment of non-condensables in the water as it flows down the tail leg. When the noncondensables reach the surface of the water, they collapse, because the pressure is higher there than in the main barometric condenser. The released energy violently excites the tail leg, causing the problems.

We reasoned that, if the surface could be raised to the area of a solid support, the energy-induced vibration might be contained. A level 13 ft above the water surface was chosen, because it gave a margin for errors in calculations, and because it cut chances of flooding the main barometric condenser. A 15-in. orifice was selected, fabricated of mild steel, and epoxy-coated to prevent corrosion.

Results were good. Over the entire range of flow, the violent vibration of the tail leg was much less. As flow increased above 10,000 gpm, the vibration from collapse of entrapped noncondensables ceased.

D F MADURA, *Mystic, Conn*

Cavitation can be culprit, causing pulsations

BD's problem suggests that self-excited vibrations are present. This means that the vibrations are sustained and amplified, possibly leading to resonance. These dynamics are frequently present with fluid motion, as in liquids, vapors, and gases. The forces maintaining vibration are controlled by the motion itself. As velocity of the motion varies, the force magnitude also varies, disappearing when motion ceases.

The system vibrates at its natural frequency, independently of the forcing frequency. Natural frequency of a fluid conductor usually becomes less as fluid velocity increases. This is because of higher deflections caused by increased inertial forces (impulse/momentum relation). This is particularly important for flow in thin-wall pipes and ducts.

Because pressure inside a primary barometric condenser is below atmospheric, liquid inside the tail leg will tend to flash, causing cavitation that leads to pulsations. Installation of a flow-resistive element (orifice) inside the pipe, near its outlet, will provide added Δp downstream of the condenser, decreasing the pressure difference between condenser and pipe outlet (sketch, 8.4B). This assumes a condenser pressure of 2 in. Hg abs, a 20,000-gpm flow

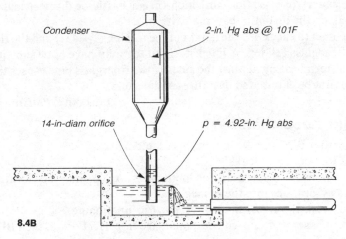

Condenser

2-in. Hg abs @ 101F

14-in-diam orifice

p = 4.92-in. Hg abs

8.4B

rate, and an orifice diameter of 14 in., with a beta ratio of $d/D = 0.25$. If BD knows the actual primary condenser pressure and tail-leg ID, he can select an orifice diameter to satisfy his actual conditions.

H B Wayne, *Jamaica, NY*

Try tail-leg changes to overcome siphoning

Vibrations in the tail leg of a barometric condenser are generally started by a high cross-sectional velocity, which creates a siphoning effect that entrains vapor and causes pulsations. Self-venting flow will occur within these limits:

$$0.31 < V_L \div (g_L D)^{0.5} < 1$$

where V_L is the cross-sectional velocity in ft/sec, g_L equals 32.2, and D is the pipe's inside diameter, in feet.

At the rate of 20,000 gpm, the vibrations are probably flow-induced. To eliminate the problem, here are a few ideas:

- Make the tail leg 34 ft long and as straight as possible.
- Make the condenser outlet the same size as the tail leg.
- Set the tail leg at least 2 ft above the hotwell bottom and 1 ft from the side.

■ Make the volume of the hotwell behind the weir at least 1½ times that of the tail leg.

■ Keep a 1-ft minimum seal on the weir in operation.

D B SYLVIA, *Roswell, Ga*

Lining may be necessary, after finding source

BD should first find out if the pipe is the source of vibration or if the vibration is the result of a resonance of some other disturbance. Identify the vibration pattern. Temporary restraints in strategic locations may help identify areas in which permanent pipe restraints should be installed. Check to make sure that the barometric condenser, or the flow pattern from it and through the tail leg, is not the source of vibration, in which case a baffle or diverter installation could break up the turbidity-causing pattern.

Other points to check are the height of the tail leg (>34 ft) and the flexibility of the fiberglass tail leg. A lined steel tail leg may have to be substituted if the flow pattern beating against the sides of the fiberglass pipe causes the flexing and cannot be eliminated by baffles or diverters.

J E JANDA, *Pottstown, Penn*

Constant-head chamber?

Some steps that BD can take to solve his problem are:

■ Look for condenser leaks.

■ Check condenser internals for loose parts. A part may be vibrating and transferring vibrations into the tail leg.

■ Check pumps, valves, and sprays, too, for functioning.

If the condenser and related equipment are in perfect order, then BD could consider installing a leg in a supply pipe or in the tail leg to act as a constant-head chamber or fluid-line shock absorber.

C R KLEIN, *Wauwatosa, Wis*

8.5 Rate intangibles in evaluating bids

Our petrochemical installation is beginning an expansion, plus some upgrading of existing equipment. The company's Central Engineering Dept, in which I work, is doing much of the engineering for the projects. I have been assigned to work with purchasing on getting bids for gas turbines, pumps, compressors, valves, and many other components. Although collecting, evaluating, and coordinating specifications is tiring and exacting work, I can understand most of it, in the light of my engineering experience. What puzzles me is the bid-evaluation factors of our purchasing section—factors evidently developed over the years in cooperation with central engineering.

We evaluate such factors as price, efficiency (where applicable), and

operating costs to get the estimated total cost over the life of the equipment. This I can understand, even though some of the past projections
went badly astray because of fuel-price increases. Some secondary factors are not so easy to understand, however. Reliability gets a certain
percentage, varying for different classes of equipment, but a vendor's
general reputation and our company's experience with his equipment
also get percentages. It seems to me that much of this depends on the
opinions of engineers and supervisors. Besides, there seems to be
redundancy for these intangibles. Experience appears to overlap with
reliability, for one thing.

Can POWER readers tell me what their experience has been with evaluation of intangibles in buying machinery and equipment? What is the best
and fairest way to take account of reliability and experience?—CSL

These seven ideas can help before buying

We are doing exactly the kind of work that CSL is, and can offer him some
help. Here are seven elements of comparison:

■ Supplier reliability must be considered. Often a proposal may be tech-
nically correct and attractively priced—but the supplier may be quoting on a
unit he has never built before. Make sure you find out where each supplier has
similar units in operation, how long they have been installed, and what operating personnel can be called for information.

■ Service from the supplier is crucial for some items, such as turbine/
generators and boilers. Ask each supplier what service facilities he has, their
location, number of men employed at the facility, setup for emergency work,
etc.

■ Expected life of the proposed equipment needs a check. A change in
materials may double the life. In any case, past performance at other plants
can indicate equipment life.

■ Price intangibles can concern firmness and the possibility of escalation.
For escalation, the formula and the maximum possible escalation are important. Payment terms have large impact, especially with expensive equipment.
In general, orders should be placed with as low a down payment as possible. In
addition, a percentage of the price should be withheld until the equipment is in
operation.

■ Delivery considerations can mean that, if equipment is needed in a hurry,
the supplier with the shortest delivery time should get priority. If equipment is
not needed until other related equipment is received, then it is much better to
accept for later delivery, with late payment of most of the cost.

■ Spare parts should be included in the equipment comparison. Depending
on the equipment, suppliers should be asked to quote on recommended spares
for several years. Costs of spares is always much lower if they are purchased at
the same time as the main equipment.

■ The equipment operators sometimes must be taken into consideration.

Mobile equipment especially is in this category. Operators usually have preferences for given brands. If another is purchased, they sometimes sabotage the unit simply to prove that they were right in their recommendation.

There is no fixed rule covering all equipment. Final decision must rest with a person experienced enought to evaluate the intangibles.

S Martinez, *Flushing, NY*

Rate intangibles on a sheet from 'excellent' to 'poor'

Competent buyers have a rating sheet for each vendor, showing ratings varying from "excellent" to "poor" for price, reliability, integrity, quality, service, and reputation. The ratings are based on feedback from the engineering and production forces, plus some involvement by the purchasing staff. Price is of primary concern, although first cost can be secondary. Reliability includes performance on vendor delivery promises. Some vendors make promises to get an order, then fail to meet the date. Is the vendor's advertising believable, and is his product line as inclusive as he intimates?

Integrity involves decision on whether the price is firm, or whether the vendor will want more money for "unforeseen developments." Will the vendor make good on equipment failure? Has he misrepresented anything about the product? Evaluate reputation by speaking to other customers of any new vendor. A good firm will supply references if you ask.

B B Brown, *Hagerstown, Md*

Beware of 'overall' reputation

Reliability can apply to two different aspects of equipment procurement and performance. Reliability of the vendor company means that it can meet promises, such as those for a delivery schedule. Also that it will accept a penalty clause in the purchase contract. Avoid the term "as soon as possible," just as you would refuse a bid at the price "as low as possible."

Reliability of equipment over a period of years is the second aspect. A vendor's general or overall reputation is virtually meaningless to you. It is his reputation in past relations with your company that is the best basis for decision. In general, too, it can make for mutual understanding if your vendor is also a customer.

F G Norris, *Steubenville, Ohio*

Try a simple point system to make comparisons

In a choice among several pieces of equipment, an engineer has three areas in which he can make comparisons: price, performance, and intangibles. In pumps, for example, performance can be thought of as guaranteed efficiency, while intangibles can cover many items that affect relative reliability. Shaft span and shaft diameter, which determine shaft deflection, are two. Internal clearances tie in with shaft deflection to give assurance against contact at close-clearance joints. Freedom from stress caused by thermal expansion or thermal shocks, quality of hydraulic thrust-balancing, and the competence of design and application engineers are included.

The two characteristics of high efficiency and utmost reliability are fundamentally incompatible in a pump. Equally unfortunate: Although the comparison areas of price and efficiency lend thmselves to direct numerical comparison, intangibles do not. Because of his professional training, an engineer is discouraged from giving serious weight to evidence other than what can be expressed numerically. This is regrettable, because there will always be choices for which valid comparisons must include evaluation of intangibles.

Artificial numbering systems have been developed for evaluating intangibles, but the practice is rare. More frequently, engineers either give weight to intangibles without trying to justify themselves, or they ignore the intangibles entirely. For as logical an approach as any that I have seen to a numerical rating system, here is one advanced by a friend:

Of a total of 100 points, reserve 40 for the price of equipment, and score this so any equipment costing double the lowest bid scores nothing on price. Divide the remaining 60 points on an arbitrary basis, such as 10 for manufacturer's reputation, 10 for buyer's experience with the equipment, 10 for the industry's experience with the equipment, and 10 or 15 for engineering prejudice. The rest of the 60 points can go to such elements as reliability and likelihood of failure, where failure may mean heavy damage to equipment, public, or employees. Compare the bids on this basis, and select the equipment having the highest total. From experience, when the results of such a point score are close, it does not matter which of several alternatives is selected. On the other hand, where major differences in the point scores occur, no amount of argument will change the score enough to affect the choice.

I J Karassik, *Mountainside, NJ*

Translate to life-cycle costs

From procurement's point of view, there should be no intangibles. The examples that CSL mentions should be translated into standard life-cycle cost parameters. From reliability of product and vendor experience, calculate mean time between failures (MTBF), availability figures, level of repair decisions, and attendant costs. Vendor dependability and reputation for service response translate into expected costs of late delivery, service delay, etc. "Intangibles" are part of the evaluation. Central Engineering has turned vendor product intangibles into performance costs, for procurement analysis.

W D Spencer, *Virginia Beach, Va*

8.6 Ways to prevent lube contamination

Our textile mill seems to have a lubricant problem that may be common in other plants or industries. We have given considerable effort to lubrication programs and training of lubrication men, until we now have what I believe to be an efficient and competent system. Nevertheless, we still get bearing and seal failures that I trace to contaminants in the lubricants. Plant dirt, water, fibers, even acids and alkalis have been identified when machinery has gone down. Besides that, I see considerable contamination in oil and grease from machines that are still running satisfactorily.

Lubricant suppliers whom we have asked for advice have given us a long list of steps that we can take. Extensive filter installation, periodic reworking of oils, a large filter-inspection program, and expensive modifications to our oil reservoirs are among the recommendations. I would like to know which steps have been found most practical in other plants. Recently, I heard of a portable cart that is taken to a machine to refilter the oil in the sump. Maybe we could benefit from ideas like this. Have POWER readers worked with equipment like that, or with other methods of preventing and removing contamination from lubricants? Is contamination a problem that can be solved?—REN

Filter cart supplies makeup oil to reservoirs

My experience with numerically controlled machine tools, which require high-purity oil, suggests several pointers for REN. First, portable filtration is a must. A cart or filter buggy, with portable pump and filter unit, supplies all make-up oil to the reservoirs, and filters reservoir oil regularly. Connections are industrial quick-connect hydraulic couplers. Fittings locations on the reservoir should create a flow in the reservoir and prevent short-circuiting. If types of oil vary widely, flush the filter-buggy before changing. If oil samples contain water and plant dirt, check tank covers and vents to see if that's how dirt or water is getting in. Water can come from cleanup hoses or from condensation during a shutdown that allows oil temperature to vary greatly.

M L REMBIS, *Mt Orab, Ohio*

Oil room should be separate and clean

Bearing seals may be the key to REN's troubles. Worn seals will allow contaminants to enter bearings. Another cause of contamination is inadequately sized vents on lube-oil reservoirs. Properly sized vents permit water to evaporate, cutting down on water-related troubles. All vents should have filter-type covers to keep out foreign matter, of course.

REN should also set up an oil room, in a clean environment, not too hot or too cold. Keep lubricants and lubricant-handling equipment here. Provide separate covered and clearly labeled containers for various lubricants. Don't use any dispensing device for more than one lubricant. And regularly clean all tanks, containers, and handling equipment.

W C McCARTHY, *Brant Rock, Mass*

Filtration of oil through media is best method

Filtration of oil through media that will remove the finest particulate matter present is the best way to take out these contaminants. At present, at least two companies make filtration units able to remove submicronic matter and water. These units can act as polishing filters in sump systems. In nonsump systems, they can be in bypasses. A double treatment will prevent acid formation. Water removal and neutralizing discs of magnesium are the steps. Regular lab analysis of the oil is also necessary, to determine when to change filters.

J A GAY, *Idaho Falls, Ida*

Trouble may be airborne contamination

Perhaps dirt, and such substances as acids and alkalis, are blowing into bearings at REN's plant. For example, if a bearing is close to a fan or blower casing that lacks sufficient sealing at the shaft on the pressure side, escaping air may be blowing into bearings and carrying grit from the outside air into the bearing. A simple two-piece doughnut ring clamped to the shaft will usually deflect the escaping air.

Another possible source of contamination is a chemical reaction between piping and oil. If the oil contains sulfur and the piping is galvanized, zinc oxide may form. Scale and rust are other contaminants, which pickling before start-up can remove. Water is another problem, especially where temperatures don't stay high enough to prevent condensation. Electric heaters or a nitrogen purge is a solution. Switching oils is not recommended, because combining the chemical additives could result in sludge. If a switch is necessary, flush the entire system before filling with new oil. Check with the oil manufacturer before replacement.

R M CORTELLINI, *Bristol, Penn*

Clean up the storage of oil and grease

The most common cause of lubricant harm is careless storage, resulting in contamination from water, dirt, and other impurities. Store oil and grease in a clean dry room set aside for this only, and heated in the winter. Don't store oil and grease near boilers, steam pipes, or heaters—overheating of grease can change its physical characteristics.

Transfer oil to storage tanks by pump. Oil-storage tanks should be raised off the floor and should carry a bottom drain valve. The vent breather line should vent to atmosphere, with an elbow and flame screen facing down. The oil-removal pipe should terminate in the tank slightly above the lowest gage-glass cock, so all oil trapped will be clean.

With grease, special covers are available for the standard 5-gal containers and 30-gal drums. The covers carry hand-operated pumps to make grease-gun filling easier. The portable cleaning rig that REN mentions is excellent for closed lube systems, such as for turbines and large boiler-feed pumps. If equipment gets lubrication from a central oil system, a permanent centrifuge could be advisable. And if the system has lube-oil coolers, be sure to clean them at shutdown, on both oil and water sides.

D Duvo, *Suffern, NY*

Recommends caution on portable filter

The portable-cart idea that REN mentions should be approached with caution if the proposed operation is in an industrial plant. We looked into the idea some time ago, after hearing that it had worked well in some critical high-technology applications. What we found from several users was that the periodic filtration of oil in severe service did not have much effect on bearing and seal life. In many industries, contaminants work into the bearing housings from outside, and stay in the bearing area until the damage is done. The sweep of oil circulation, going back to either the regular sump and filter or the portable-cart filter, is not enough to clear out all the oil from pockets and surfaces. Result is damage rates that are fairly constant.

R W Kolchak, *Chicago, Ill*

Differentiate between two areas

Consider two areas in lubricant contamination: dirt in the lubricants before application, and contamination in the equipment. Never expose open lubricant containers to a dirty environment. As the oil leaves a barrel, for example, the contaminated air will be sucked into the barrel. A filter on the barrel vent will correct this difficulty. Keep oil and grease drum heads wiped down and dry. Wipe funnels dry with lintfree material. Fill grease guns in a location outside the dirty atmosphere.

Preventing lubricant from being contaminated in the equipment requires a study of the lubrication system of each machine. If a machine has a gear reducer, remember that the reducer breathes as temperature changes. Most reducers have a filtered breather that seldom is cleaned or changed. Get filtered vents on reducers and clean them regularly.

Cover oil sumps, and even install an instrument-quality air supply to maintain slight positive pressure. Shields over pillow blocks will reduce exposure to airborne dust and give some protection against water. Much manufacturing machinery is too complicated to permit all-inclusive lube protection, but REN should do whatever is possible. For example, the shop mechanics could build a cart with pump, filter, and water separator on it, and this might extend the life of the oil.

B B Brown, *Hagerstown, Md*

8.7 How can lubricant costs be reduced?

Our engineering department has general supervision over several plants in diversified process and manufacturing industries. Up to now, we have let each plant go its own way in buying such supplies as lubricants, hydraulic fluids, and cutting oils. Now, however, we are starting to realize that purchases of these items add up to sizable amounts in a year. Prices keep going up, too. Our plants have always paid close attention to lubrication, with programs carefully worked out and results periodically reviewed. We have no complaints on that score.

Our aim now is to save ourselves some money on oils, greases, hydraulic fluids, and cutting oils. We are hesitant to some extent because we have always followed vendor advice, buying premium products and throwing away the lubricant when it showed any impairment. I think we want to look at reclaiming and reprocessing methods, along with ways to extend lubricant and fluid life.

Can Power readers tell us what steps to take to cut our purchases while still protecting our machinery? Can our small plants reclaim their lubricants, or is this only for large plants? And how do we convince our maintenance supervisors that reclaimed and reprocessed lubricants are as good as the high-grade new oils and fluids to which they have been accustomed?—MR

Make full use of lube-oil improvements

We have found that vendors are possibly not the best sources of advice for determining periods of effective use from oils. Our experience shows that the lives of some lubricants are underestimated by the manufacturer. Also, especially on older machines, the lube-oil renewal schedule that MR is using may not recognize the superiority of today's lubricants. Consult manufacturers of equipment for lengthened oil-change intervals, naming the brand and type of oil. This suggestion may halve oil consumption. Other recommendations we can make:

■ Care for the oil before and in use. Cooling and filtering in service prolongs life, and centrifuging restores cleanliness.

■ Use oil twice. Oil drained from sumps and gearboxes can go to small hand-pump lubricators or on gears and chain drives that are subject to rapid contamination anyway.

■ Burn the oil. This can save on fuel and disposal carting, but make sure local codes don't prohibit the practice.

T A Riso, *Richmond Hill, NY*

Synthetics will help compressors

Re-refined oil is certainly reliable if it is processed in a good place. If the oil must meet detergent specs, or any other specs, the re-refiner can put in the additives and mark the oil with spec numbers. The re-refiner is also a good place to send any dirty or used oil that a plant wants to discard.

For air compressors, synthetic lubricants are excellent, on the basis of our experience. In one case, a compressor has run untouched for over six months, quite a different performance from the previous 100 hr tops per oil change.

G ROSEKILLY, *San Mateo, Calif*

Don't lose sight of overall cost

MR's interest seems to lie in *lubricant-cost* savings, rather than in the more fruitful and less risky field of *lubrication-cost* savings. In the latter area, especially in these days of high downtime and labor costs, more-expensive lubricants save money. For example, an extension of only 20% in an oil's drain interval frequently can justify a threefold increase in the lubricant's unit cost.

If lubricant cost is what MR is interested in, his engineering department should inventory all lubricant-consuming equipment in the company and then, aided by the lubricating-oil vendor or by a lubrication manual, such as US Steel Corp's, determine the minimum number of lubricants that will satisfy all the equipment. Bulk purchases, centralized lube systems, and specialized dispensing equipment are three possibilities.

MR should not throw away lubricants—this is costly, and is also illegal in the US. Frequently, a hydraulic oil or circulating oil is changed before it has "worn out," because of the equipment maker's recommended drain interval. Examine this oil. If its color and odor are good, filter the oil through a 10-micron industrial filter and store it. Then use it in systems that consume oil, or in noncritical oil-can service.

If the plant generates substantial amounts of used oil, speak to local oil blenders or industrial-oil re-refiners about reworking possibilities. *Keep used oil segregated.* It makes the re-refiner's job easier and lowers reworking cost. MR's staff should not try to reclaim or re-refine their own oil. Oil re-refined by a competent firm may be even better than new.

C L PATTISON, *Frenchtown, NJ*

Order year's supply at a time

The engineering department should take inventory of all lubricants required in a year by the plants. Then tabulate the quantities and get prices from several oil companies on the total. After obtaining bids, the department can order a year's supply for each plant. When the lubricants are delivered, the trade names may not be those the maintenance men are familiar with, but the material will be the same. Reclaiming? Only for cutting oil; burn the rest of the used oil.

G H HILL, *Sunbury, Penn*

Heating and filtering works, too

We operate over 25 cold-storage warehouses across the country, and are very successfully using re-refined refrigeration-compressor oil. Since installation of an oil-filtration system in one plant in 1971, we have cut our oil purchases from about 660 gal/yr to 55. These savings are typical for our plants on re-refined refrigeration oil. There seems to be a hidden economy in this, too, in lubricating bearings and blades of our rotary booster compressors, where oil can now be applied in liberal amounts with the assurance that it can be recovered and re-refined. Blade wear seems to be reduced.

Our system, developed at our Omaha plant, includes oil withdrawal, heating and filtering, and storage and return to the system. In the second phase, a non-Code 30-gal heating tank with electric heater delivers to a 1-gpm rotary gear pump. Heating is to 215F, to purge water and volatiles. Oil is pumped through two three-stack filters, and back to the first tank, until clarity is adequate. In addition to the usual inspections by our engineers, we send samples to a test lab periodically. Unobjectionable viscosity increase is the only change.

C A TOOGOOD, *Lyons, Ill*

Plant builds its own filters

Oil reclamation is feasible in large and small plants. Most oil is discarded because it is contaminated by water or dirt. Heat takes out the water, but filtering or settling removes the dirt. This year, we fabricated two drum reclaim filters, as the sketch shows. These were suggested by our lube supplier as a try

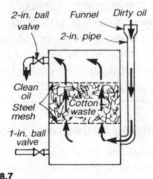

2-in. ball valve · Funnel · Dirty oil · 2-in. pipe · Clean oil · Steel mesh · Cotton waste · 1-in. ball valve

8.7

at reclaiming lubricating and hydraulic oil. To date, we have averaged reclamation of two drums a month. Some oil can't be reclaimed, because it won't pass lab tests performed by our supplier after we filter the oil. We have been able to reuse 70% of the filtered oil, however—a good average for a minimal investment. Just change the cotton waste as required.

M W HARTENBERGER, *Rockford, Ill*

Ultrapurification protects bearings

MR can extend fluid life by reprocessing, and he can also extend equipment life and cut downtime. Tests on effects of submicron contamination on bearing life have demonstrated that ultrapurification extends bearing life dramatically. Quite possibly the saving in fluid and the reduction in machine-repair costs will quickly pay out.

Equipment is available that can remove all particulate contamination down to fractions of a micron—without taking out corrosion inhibitors or other dissolved additives. Skid-mounting the equipment allows one unit to serve several plants.

C F Humphreys, *Torrance, Calif*

8.8 Secondary explosions outside engines

Our close involvement with diesel engines has brought us some first-hand knowledge of crankcase explosions in these machines. In several of the cases that have come to our notice, there has been damage outside the engine from what appears to be a secondary explosion. Windows have been blown out and paint burned on walls as far as 100 ft from the engine itself. Strangely enough, the internals of the engine did not seem damaged nearly as much as the outside area. This leads us to believe that, if a secondary explosion did occur, it must have been outside the engine, rather than in the crankcase where the first explosion took place. Because this is, to some extent, contrary to what most operators believe, we would like to get the opinion of Power readers on the causes. The phenomena we have seen apply to both oil and gas engines.

Have Power readers noticed this kind of secondary explosion during operation of oil or gas engines? What would be a reasonable means of protecting a plant against it, in view of the fact that primary explosions occur in many engine installations?—DNL

Vapors around engine cause explosions

At times, flammable gases are present outside engines, and all that is needed for an explosion is something to ignite them. Some engine or mechanical rooms have enough gasoline or oily vapors present in a tightly sealed area to allow this to happen. The vapors cannot readily escape through openings in the building or through windows, nor can fresh air enter through those openings. So, once a primary explosion occurs, it sets off the explosive medium surrounding the engine and creates a secondary explosion. In an area with solid walls, such as brick, the initial shock waves reverberate off the walls and break windows, or cause fires and charring.

J Rampolla, *New York, NY*

Keep room well ventilated to stop reignition

The volume in the crankcase zone of most diesels seems to me to be too small to be able to produce a serious explosion outside the engine after a primary explosion inside the crankcase has consumed the initial quantity of oil vapor. The fact that a secondary explosion often occurs inside the crankcase would seem to support this idea. The air entering the engine crankcase after the first explosion mixes with the newly vaporized oil, and the mixture reignites, but this occurs inside.

Window breakage outward can result from a shock wave originating at an engine vent, and is not necessarily the result of an external explosion. To burn wall paint over a wide area would require a large outside volume of combustible material, and it is hard to believe that the amount of oil escaping could mix thoroughly enough and quickly enough with the engine-room air to set up an immediate explosion. I suggest that external vapor concentrations already in the engine-room space are being ignited by flames from an internal explosion. If this is the case, then the best way to prevent the outside explosion would be to keep the room well ventilated.

G H BALLARD, *Houston, Tex*

Venting explosions can be unsafe, dangerous

Crankcase explosions are the result of oil mist or vapor, plus air, plus a source of ignition. I spent a number of years with large marine diesels and experienced a number of explosions, but I never heard of an explosion unless the above prerequisites were met. Oil mist will not explode if it is too diffuse, nor if it is too dense; it must be between the lower and upper explosive levels. Generally, a hot bearing creates the necessary conditions, first producing the vapor and then igniting it. DNL appears to be venting his explosions, which, although dangerous, is one method of dealing with them. It is better, however, to detect the incipient condition by the use of suitable instrumentation.

E FRANCIS, *Bowmanville, Ont, Canada*

Oil mist blown out is cause

Perhaps my experience in tests on experimental explosion chambers will help DNL. We have two of these chambers, one with a volume of 14 ft³ and the other with 91 ft³. The volume of each can be adjusted with lube oil to meet a wide range of test requirements. The oil can be stirred, while being maintained at 150F, and an oil spray, also at 150F, can be introduced. A fan creates turbulence in a mixture of natural gas and air, which is then ignited by a spark. These conditions duplicate those in the crankcase of a typical operating diesel or gas engine.

During several tests, secondary explosions occurred outside the test chambers. Several explosions were of considerable magnitude. In one case, I was standing 33 ft at right angles to the 91-ft³ combustion chamber, and the secondary explosion in the atmosphere produced flame travel that caused burns on my arms and face.

A review of pressure recorded in the test chambers with various mixtures of natural gas and air indicated that maximum pressures resulted when the gas/air ratio was approximately 8.5%, and the explosion range was from 5% to 14% by volume. With a crankcase explosion where a relief valve has no internal flame trap, a considerable amount of oil mist comes out to the atmosphere along with the flames. If the amount of oil-mist discharge is right, the secondary explosion will occur. Intensity will vary with the resulting air/oil-mist ratio of engine-room atmosphere.

In an engine room measuring 40 ft × 40 ft × 20 ft, the volume exposed to explosion is 32,000 ft^3. Even if only 10% of this volume becomes contaminated with fuel and oil mist mixture and is ignited by the flame emerging with the original explosion or by some external ignition, it is evident that the secondary explosion will be large and forceful. I believe that oil-wetted flame-arresting screens and sufficient area of relief valves to limit explosion pressure to about 30 psig are good measures.

H LEACH, *Troy, Penn*

Alert engine operators are good defense

A crankcase explosion is a serious accident. All engine operators should be aware of the causes, and should remain alert to detect possible symptoms. One precaution to prevent damage outside the engine is to keep the engine-room air well below the lower explosive limit for oil vapor or gas. Then if a crankcase explosion should release more oil vapor or gas, there will be less chance of an explosion outside. As far as release of vapor from the engine is concerned, perhaps it is better for some of the inflammable material to be ejected from the engine, and thus reduce the amount remaining inside for a later explosion.

R E HALSTEAD, *Minneapolis, Minn*

Cooperative research is only way to go

DNL's experiences are another good reason why we should try to find out more about the nature and exact time sequence of events during an engine explosion. The necessary knowledge will never come from post-explosion questioning of operators, because the events occur in milliseconds. Analysis and investigation of bearings, crankcase scavenging, engine parts and building components will always be useful, of course, but even this cannot take the place of scientifically controlled and reproducible experiments at full or nearly full scale. I would like to see some cooperative research on this by our engine manufacturers, insurance companies and government agencies (such as OSHA and DOD).

A V MARCUS, *New York, NY*

8.9 Plant air conditioning at low cost

At our southern Ohio location we manufacture electrical control equipment. Our owner has seen several completely air-conditioned plants and now wants to air condition our operations. It is my opinion that the idea is still fairly new and that unless we proceed with caution we will waste money. In addition to talking with consultants and other plants, I would like to get Power readers' thoughts on how to hold down the cost. We have about 60,000 sq ft in two buildings that house presses, automatic screw machines, finishing operations and some heat-treat work. One building, 60 years old, has many windows and skylights. The other building is new, with a flat roof.

Some of the ideas that have occurred to us so far include closing all windows and skylights and adding to our lighting. This will raise our electrically produced load, we realize. Another idea is use of spot air conditioning at worker stations, with the area either unenclosed or partly enclosed. This idea has some obvious drawbacks, too. Based on experience, what do Power readers think about these two approaches? Can your readers give us any other help in planning and installing a total plant air conditioning system?—LA

False ceiling could cut volume of air to be conditioned

One way to hold down the cost of air conditioning the plant would be to transfer operations that do not require air conditioning to a single building. If that is not possible, then a centralized chilled-water system for both buildings, with portable walls separating each into sections, would be a solution. This would permit utilization of the a/c areas to increase working efficiency.

Painting windows and skylights of old buildings with aluminum paint would reflect heat and also reduce heat transfer. Another possibility is a false ceiling in the old building, to reduce the total volume to be air conditioned.

T W Hammond, *Los Banos, Calif*

Exhaust the heat from lights to cut costs

LA's ideas for air conditioning his plant are satisfactory, except that it would not be advisable to employ the spot-cooling system. Physiological effects vary from individual to individual, and an air current that can be felt, or a draft, may result in head colds. Why not isolate the high heat producing areas from the general air-conditioned area and also cool the operators' rest or break area?

LA should not stint in the closing up of leaks and use of thermal insulation. Air-conditioning operating costs are high; our plant experience indicates 35 to 40% increase in electric power costs during the cooling season. A closed system with return-air loop should be used to hold down operating costs. Double-door setups at personnel entrances and air curtains at large doorways used for materials handling will help.

When the windows and skylights are blocked, the increased lighting load can be held to minimum by use of troffered ceiling fixtures with an exhaust system for the heat generated in and over the dropped-ceiling area. Use a light-colored flat paint or enamel on the walls to get full lighting value without the glare that can come from glossy walls.

If LA goes to dropped ceilings and troffers, he can reduce operating costs significantly if the panels are perforated and the exhaust air flow reversed in winter. An office building near our plant gets enough heat from this system until the outside temperature drops into the 20's.

B B Brown, *Hagerstown, Md*

Consider putting the roof to work

Even though LA's plant will probably never know whether it has gotten its investment back in better productivity and improved product, economics is still important in deciding how to air condition the plant. Most important guideline is to base the size of equipment on minimum conditions; this will keep initial and operating costs down, and the effect will be nearly as welcome as if the equipment had been designed for the hottest day in southern Ohio.

From what LA says, I assume that his steam-boiler plant is not big, so that electric drive would be his best choice. If he has single-story buildings, the roof could easily be his best spot for setting air-conditioning equipment. Air should be discharged at the 10-12 ft. level. If ceiling is high, watch the ductwork costs. If LA has multistory buildings, rooftop mounting may require expensive floor cutting and ducting, so that inside location of equipment could prove cheaper. It requires more floor space, but then LA's management may not be able to put a close figure on what the area is worth.

R D Farbstein, *Chicago, Ill*

Spot a/c works but may have high upkeep

Closing of windows and skylights does not usually increase the lighting load. For example, when plant operates at night, the lighting levels will be unchanged. Same is true for overcast periods and during rain or snow.

Getting down to the air-conditioning system itself: while un-enclosed or partly enclosed spot air conditioning may be cheaper initially, its operating expense may be higher because of heating and cooling loss to adjacent areas. For lowest total annual cost, including fixed charges and operating expense, LA should look into conditioning each area to satisfy minimum ambient conditions based on work performed.

H B Wayne, *Jamaica, NY*

Tangible payoff? Reader doubts it

Lately I have been hearing that installation of air conditioning in old industrial buildings is running up to $3/ft^2, and if LA meets this kind of cost, he will be paying about $180,000 for the work. This will be followed by constant maintenance charges and then replacement costs for equipment. He should also remember that if the company moves, it cannot very easily take the a/c system with it in the way it can the machine tools and process equipment. Nor will the new buyers give him full price for the system, because few firms think they need full air conditioning.

Nevertheless, here are two possible pointers for his proposed system. First, spot air conditioning is not good; the workers tend to congregate, and there is a health problem when workers go from cool to hot areas frequently. Ordinary fans are best for high-temperature areas with dry heat.

Second, it might be possible in some zones of the plant to close up the zone, allowing access through air locks with double doors, and retain the coolest air of the preceding night period. This has been tried occasionally in some warehousing operations, where space is large and only a few workers are present. The zone's fans bring in the cool air at the crucial time, and are then shut off until the zone has warmed up to the point at which they are again needed.

G R WALLACE, *St. Louis, Mo*

Aluminum paint will aid LA

A simple and low-cost solution would be spot air conditioning plus translucent paint to reflect solar heat rays but allow light to enter. To maintain a comfortable temperature near the floor, locate the discharge of the spot air conditioners at about ten feet from the floor. Then close all openings below this level as well as possible. Reflecting aluminum paint on skylights, windows and even glass doors will effectively reflect heat but pass light.

C R KLEIN, *Wauwatosa, Wis*

Index